SOTHIS PRESS

Sothis Press aims to completely reimagine the public domain book with an emphasis on universal access to captivatingly designed print editions. We believe that technology will enable every single book ever published to be perpetually accessible in beautiful print form. Sothis is beauty in ubiquity.

Printed and bound by CPI Group (UK) Ltd, Croydon, CR0 4YY

www.sothis.press
1 Waterhouse Square, London, EC1N 2ST

Rudimentary Treatise on Water Power

Driving Flour Mills and Hydrostatic Engines

Joseph Glynn

SOTHIS PRESS

ISBN 978-1-0373-1105-5

Rudimentary Treatise on the Power of Water as Applied to Drive Flour Mills and to Give Motion to Turbines and Other Hydrostatic Engines is in the public domain.

EDUCATIONAL SERIES.

e following are the Works already published in

CIVIL ENGINEERING, &c.

e Volumes are bound in limp cloth, except where otherwise stated

CIVIL ENGINEERING, the Rudiments of. By HENR LAW, M. Inst. C.E. Including a Treatise on Hydraulic Eng neering. By G. R. BURNELL, M. Inst. C.E. Seventh Editio revised, with Large Additions on Recent Practice in Civ Engineering. By D. KINNEAR CLARK, M. Inst. C.E. 6s. 6d. cloth boards, 7s. 6d.

THE CONSTRUCTION OF GAS-WORKS, AN THE MANUFACTURE AND DISTRIBUTION OF COA GAS. Originally written by SAMUEL HUGHES, C.E. Sevent Edition. Re-written and much enlarged by WILLIAM RICHARD C.E. With numerous Illustrations. 5s. 6d.; cloth board 6s.

WATER-WORKS, for the Supply of Cities and Town. With a Description of the principal Geological Formations England as influencing Supplies of Water; Details of Engin and Pumping Machinery for Raising Water. By SAMUEL HUGHE F.G.S., C.E. New Edition, revised and enlarged, with numero Illustrations. 4s.; cloth boards, 4s. 6d.

ROADS AND STREETS (THE CONSTRUCTION OF In Two Parts. I. The Art of Constructing Common Roads. I HENRY LAW, C.E. II. Recent Practice in the Construction Roads and Streets: including Pavements of Stone, Wood, a Asphalte. By D. KINNEAR CLARK, M.I.C.E. Third Editio revised. With numerous Illustrations. 4s. 6d.; cloth boards, 5

EMBANKING LANDS FROM THE SEA, the Pra tice of. Treated as a Means of Profitable Employment f Capital. By JOHN WIGGINS, F.G.S. 2s.

LAND AND ENGINEERING SURVEYING, Treatise on; with all the Modern Improvements. Arranged f the Use of Schools and Private Students; also for Practic Land Surveyors and Engineers. By T. BAKER, C.E. Ne Edition, revised by EDWARD NUGENT, C.E. Illustrated wi Plates and Diagrams. 2s.; cloth boards, 2s. 6d.

SUBTERRANEOUS SURVEYING, an Elementar and Practical Treatise on. By THOMAS FENWICK, and THOMA BAKER, C.E. Illustrated. 2s. 6d.; cloth boards, 3s.

MAGNETIC SURVEYING AND ANGULAR SUR VEYING, with Records of the Peculiarities of Needle Dis turbances. Compiled from the Results of carefully made Ex periments. By WILLIAM LINTERN, Mining and Civil Enginee d Surveyor. 2s.

ANITARY WORK IN THE SMALLER TOWNS AND IN VILLAGES. Comprising:—1. Some of the more Common Forms of Nuisance and their Remedies; 2. Drainage; 3. Water Supply. By CHARLES SLAGG, Assoc. M. Inst. C.E. Second Edition, Revised and Enlarged. 3s.; cloth boards, 3s. 6d.

VELLS AND WELL-SINKING. By JOHN GEORGE SWINDELL, A.R.I.B.A., and G. R. BURNELL, C.E. Revised Edition, with an additional Chapter, and a new Appendix on the Qualities of Water. Illustrated. 2s.

'IONEER ENGINEERING. A Treatise on the Engineering Operations connected with the Settlement of Waste Lands in New Countries. By EDWARD DOBSON, Assoc. Inst. C.E. 4s. 6d.; cloth boards, 5s.

IATERIALS AND CONSTRUCTION; a Theoretical and Practical Treatise on the Strains, Designing, and Erection of Works of Construction. By FRANCIS CAMPIN, C.E. Second Edition, revised. 3s.; cloth boards, 3s. 6d.

ICTIONARY OF TERMS used in Architecture, Building, and Engineering; Mining, Surveying, and Construction; Early and Ecclesiastical Art; the Fine Arts, &c. New Edition, revised and enlarged by ROBERT HUNT, F.R.S. 5s.; cloth boards, 6s.

RAWING AND MEASURING INSTRUMENTS. Including—I. Instruments employed in Geometrical and Mechanical Drawing, and in the Construction, Copying, and Measurement of Maps and Plans. II. Instruments used for the purposes of Accurate Measurement, and for Arithmetical Computations. By J. F. HEATHER, M.A., late of the Royal Military Academy, Woolwich, Author of "Descriptive Geometry," &c., &c. Illustrated. 1s. 6d.

PTICAL INSTRUMENTS. Including (more especially) Telescopes, Microscopes, and Apparatus for producing copies of Maps and Plans by Photography. By J. F. HEATHER, M.A. Illustrated. 1s. 6d.

'URVEYING AND ASTRONOMICAL INSTRUMENTS. Including—I. Instruments used for Determining the Geometrical Features of a portion of Ground. II. Instruments employed in Astronomical Observations. By J. F. HEATHER, M.A. Illustrated. 1s. 6d.

IATHEMATICAL INSTRUMENTS. By J. F. HEATHER, M.A. Enlarged Edition, for the most part entirely re-written. The 3 Parts, as above, in One thick Volume. With numerous Illustrations. 4s. 6d.; cloth boards, 5s.

TATICS AND DYNAMICS, the Principles and Practice of; with those of Liquids and Gases. By T. BAKER, C.E. Third Edition, revised by E. NUGENT, C.E. Illustrated. 1s. 6d.

RUDIMENTARY TREATISE ON THE

POWER OF WATER

AS APPLIED TO DRIVE FLOUR MILLS AND TO GIVE MOTION TO TURBINES AND OTHER HYDROSTATIC ENGINES

By JOSEPH GLYNN, F.R.S.

MEMBER OF THE INSTITUTE OF CIVIL ENGINEERS, ETC.; HONORARY MEMBER OF THE PHILOSOPHICAL SOCIETY, NEWCASTLE-UPON-TYNE, ETC.

Seventh Edition, with Additions and Corrections

NUMEROUS ILLUSTRATIONS

LONDON
CROSBY LOCKWOOD AND CO.
7, STATIONERS' HALL COURT, LUDGATE HILL
1885

TO THE RIGHT HONOURABLE

THE EARL OF ROSSE, K.P.

This Book

IS MOST RESPECTFULLY DEDICATED.

PREFACE.

Some of the most pleasant days of the Author's early life were passed on a small, but picturesque estate belonging to his family, on which were three corn mills—one driven by wind, and two by water. To his youthful associations with these and their tenants may in part be owed his attachment to mechanical pursuits. Subsequently his father was induced, as an investment, to build, on other property, a flour-mill driven by steam, to which, during its erection, he was a daily visitor.

In after life his professional employment as a civil engineer frequently involved the construction of water-wheels, many of them on a large scale, for drainage and irrigation, and for motive power, besides hydraulic works of various kinds.

Few persons have had occasion to use the steam-engine more extensively, and he is fully sensible of its great value and importance as a prime mover for machinery; yet he has often felt that water-power has been unduly superseded or neglected when it might have been usefully employed; there are many places where fuel is scarce, where water abounds, and where mechanical power is wanted, but much expense cannot be afforded; and if motive power be used at all, it must be obtained at light cost.

In such circumstances are some of the colonies, and many parts of Ireland: there water-power must precede the steam-engine and be the pioneer to manufacturing industry.

He has often contemplated writing some short and popular work on this subject, whenever an opportunity might present itself: this has now occurred in the publication of a series of rudimentary books by Mr. Weale; and having already written one of these, a "Rudimentary Treatise on Cranes," which has circulated very extensively, and been translated into several foreign languages, he has been induced to devote such intervals of leisure as he could obtain to the production of the present volume.

CONTENTS.

CHAPTER I.

On the Early Use of Millstones to Grind Corn, and the Employment of Water as a Moving Power.

CHAPTER II.

Nature and Properties of Water.

CHAPTER III.

On the Measurement of Effluent Water.

CHAPTER IV.

The Sources and Supply of Water; its Distribution and Use as a Motive Power.

CHAPTER V.

Rain Gauges and Reservoirs—Embankments and their Construction.

CHAPTER VI.

Horizontal Water-wheels and similar Machines.

CHAPTER VII.

Turbines and the Reaction of Water Pressure on Wheels, etc.

CHAPTER VIII.

Undershot Water-wheel.

CHAPTER IX.

Overshot Wheels.

CHAPTER X.

Breast Wheels.

CHAPTER XI.

The Water-pressure Engine and the Water-ram.

CHAPTER XII.

The Application of Water-power to Tunnelling.

CHAPTER XIII.

The Application of Water-power to Grinding Corn.

CHAPTER XIV.

The Conical Mills.

CONCLUSION.

LIST OF ILLUSTRATIONS.

RUDIMENTARY TREATISE

ON

THE POWER OF WATER

TO TURN MILLS AND GIVE MOTION TO MACHINES.

CHAPTER I.

THE EARLY USE OF MILLSTONES TO GRIND CORN, AND THE EMPLOYMENT OF WATER AS A MOVING POWER.

ALTHOUGH labour is the lot of man, and is indeed necessary to his existence, yet the human mind naturally revolts from those kinds of labour in which it takes no part, and in all ages men have endeavoured to shun mere toil involving only the employment of animal strength without the exercise of thought or dexterity. Thus it has ever been, from the earliest ages even until now, that occupations requiring nothing beyond the exertion of muscular force have been assigned to persons taken captive in war, or sold into slavery, or reduced by other misfortunes to perform constant and monotonous labour.

Such were of old the hewers of wood and the drawers of water, and when corn was first ground to make bread, the task of turning the mill was performed by the bondman or the captive; or else among rude tribes it was imposed on the women of the family while the men were engaged in war or in the chase.

Thus we read in the book of Judges, when the Philistines heaped injury and insult on their fallen foe, "they put out his eyes and bound him with fetters of brass, and he did grind in the prison-house."

So when the Prophet foretells the utter destruction and desolation of a great and mighty empire he says, "Come down, and sit in the dust, O virgin daughter of Babylon, sit on the ground: there is no throne, O daughter of the Chaldeans: for thou shalt no more be called tender and delicate. Take the millstones and grind meal."

The grinding of meal appears to have been performed in very remote ages by a pair of millstones, similar in form and in principle to those at present in use, the difference being chiefly in their size and in the mode of driving them. The Mosaic law says, "No man shall take the nether or the upper millstone to pledge: for he taketh a man's life to pledge." Such millstones were used by the Roman legions in Britain, and are often found about their ancient camps and stations, but nowhere perhaps in such number and variety as in the county of Northumberland, along the line of the Roman wall, and many fine examples are preserved by the Antiquarian Society of Newcastle-upon-Tyne.

These are generally about a foot in diameter, and made of the hard sandstone found in the coal districts, hence called millstone grit. The lower millstone was stationary; sometimes it was made larger than the upper stone, and had a rim or border to confine the meal and cause it to be delivered by a spout into a shallow vessel or sieve, but generally it was a round stone of the same size as the top stone, and must have been set on a cloth spread upon the ground to receive the meal.

An upright pin or pivot fixed in the centre of the nether millstone formed a kind of axis, and passed through a hole in the upper stone, which like the nether millstone was of a flat circular form. The upper stone was fitted with a handle to drive it round, and make it revolve over the lower stone; sometimes the upper surface of the top stone was hollowed into a shallow dish to hold a portion of the corn to be ground, which was gradually shaken down the hole round the fixed axis by the motion of the mill.

It is curious to observe that the principles on which corn-mills were constructed thousands of years ago, remain the same to the present day, and "the work" as millwrights call it, that is the system of grooves cut in the grinding faces of the millstones which crossing each other when in motion act like a pair of shears to cut the grain and also to throw it out from the centre to the circumference, are now set out and formed in the same way as they were from the beginning.

The sketch here given will perhaps afford a better explanation than many words.

Although the mill here described was a portable domestic implement, and some were even small enough to be held in the lap, yet it was often made of much larger dimensions, and was worked by many persons at once with levers or bars like a capstan, by men walking in a circle round it, and pushing

Fig. 1.—Ancient Roman Mill.

the bars so as to turn the upper stone, while in some cases asses and other cattle were employed, and eventually the Romans used the power of water to drive their mills. A writer of the time thus eloquently alludes to this important improvement:

"Cease, ye maidens, ye who laboured in the mill; sleep now, and let the birds sing to the ruddy morning; for Ceres has commanded the water nymphs to perform your task. These, obedient to her call, throw themselves upon the wheel, force round the axletree, and, by these means, the heavy mill." —*Beckmann's History of Inventions.*

The water-wheels first used to drive corn-mills were horizontal; they were of small size, and revolved rapidly. The axle passed through the centre of the lower millstone, as the spindle does now. It turned the upper millstone by means of a cross-bar fixed in the eye or centre of the stone, whilst a current of water directed against the vanes of the wheel on one side of the axle, urged it round. Such water-mills are still used in India; and a curious model of one, in its most simple and primitive form, was sent to the Great Exhibition in Hyde Park. A modern traveller informs us that "all the flour-mills upon the River Meles at Smyrna are constructed in this way, and necessarily answer well in countries where water power is abundant: their great simplicity preventing their readily getting out of repair, while costing but little also." Some of these simple mills may yet be found in remote and mountainous parts of Italy and France.

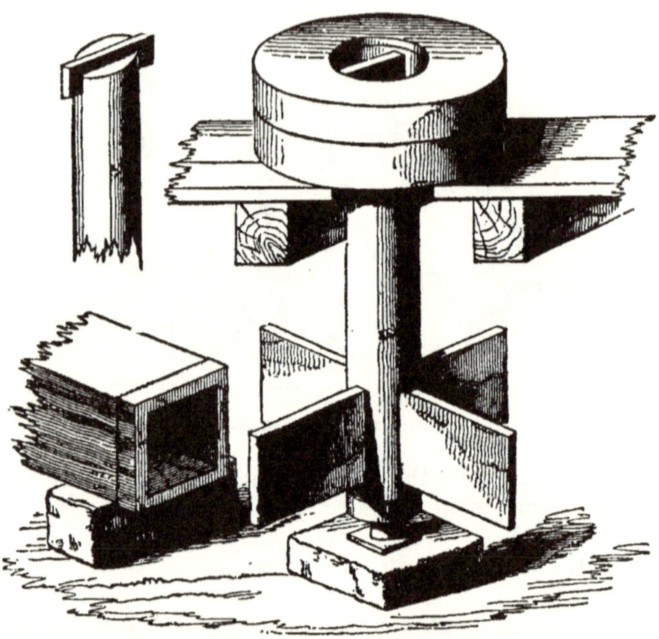

Fig. 2.—Ancient Indian Water-Mill.

The horizontal water-wheel is again coming into notice with many refinements and improvements of modern invention; and not far from the rude Indian model before mentioned might be seen the elaborate machine of Messrs. Fromont & Son, French engineers, made entirely of iron, capable of working to fifty horse-power, with a fall of two mètres, or six feet seven inches, and susceptible of such nice adjustment, as to be adapted to spin cotton and silk as well as to grind corn; while the inventor states, from his experience of similar wheels already in action, that he can obtain more than 70 per cent. of the power expended, and that, in one instance, the effective power has reached 79 per cent.

The Romans also used conical mills for grinding corn. A

very complete example of this kind was found in the excavations of Pompeii, where it had been buried for nearly seventeen centuries. The locality appeared to have been the shop of a wealthy baker; and the mill, which was of considerable size, was so fitted as to be worked by men or by cattle. It is remarkable that the conical flour-mill, in a modified form, should be again brought forward and exhibited as a modern invention, to which the proprietors attach the highest value. Two examples of conical flour-mills, both of them patented, appeared in the Great Exhibition of 1851, where, at the same time, was shown, in the East Indian section, a very neatly made conical mill, used in India for husking rice.

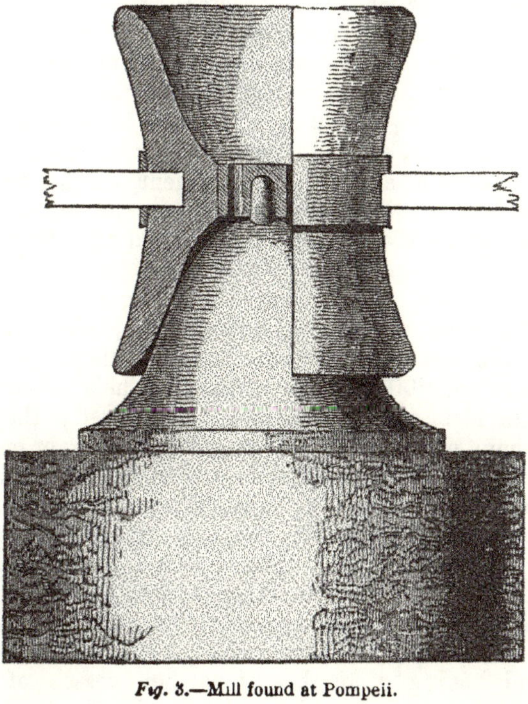

Fig. 3.—Mill found at Pompeii.

The vertical water-wheel appears to have been known to the ancients at a very early period, but it was chiefly used to raise water for the purposes of irrigation.

Examples of such wheels, in their original form and use, are still to be seen on the Nile and the Euphrates; and wheels are also employed by the Chinese to raise water for their rice fields and cane plantations; those of Egypt and

Syria generally resemble each other. The following description of them was given by Captain (now Colonel) Chesney, in his survey and report on the navigation of the Euphrates:—

"The scenery above the town of Hit, in itself very picturesque, is greatly heightened, as we are carried along the current, by the frequent recurrence, at very short intervals, of ancient irrigating aqueducts; which, owing to the windings of the river, appear in every variety of position, from the foreground of the picture to the distant part of the landscape. These beautiful specimens of art and durability literally cover both banks, and prove that the borders of the Euphrates were once thickly inhabited by a people far advanced, indeed, in the application of hydraulics to domestic purposes of the first and greatest utility, the transport of water.

"These speaking monuments of other times have, as may be supposed, suffered in various degrees during the lapse of so many ages of partial or entire neglect, and the greater portion is now, more or less, in ruins; but some have been repaired and kept up for use, either to grind corn or to irrigate, having a modern wheel attached to the ancient, simple, and most efficient model; the whole being, in some instances, sufficiently well preserved to show clearly the original application of the machinery.

"The aqueducts are of stone, firmly cemented, narrowing to about two feet or twenty inches at top, placed at right angles to the current, and carried various distances towards the interior from 200 to 1,200 yards, their height being regulated by the level of the ground to be irrigated; the shorter distances have one height of arches, and the larger ones two, one above the other, both extremely pointed: in fact, almost forming a triangle from the spring of the arch to its point. At the one extremity of the structure, which is some little distance in the river, the building makes a turn *parallel* to the stream, and there widens sufficiently to contain one, two, or three, and, occasionally, four wheels, parallel to each other, and revolving with the current, each of about 33 feet diameter, and having a number of earthen vessels of 3 or 4 inches diameter, and 20 inches long, placed at about 18 inches apart, round the exterior rim of the wheel, which is formed of light small scantling, its greatest width being rather less than the diameter of the vessels attached to it; which, dipping a few inches into the water, are filled and forced round by the current in succession, the open end foremost, until each, in turn, reaches the top, and there discharges its contents into a trough, which conveys the water from each wheel into the conduit of the aqueduct (an open one), with a gradual fall, as far as the place to be irrigated; and, as the wheels are movable, so as to elevate the axles, by means of stones or beams of wood, as the water increases, they can work equally well at any height of the river; the earthen vessels giving of themselves a sufficient impetus, without any other means, except in some few places where the current happens to be weak; in which case, six or eight fans of palm branches, each about 18 inches square, are added to the sides of the wheel's circumference, to give the water more power upon it. Such additions are, however, very rare, and the earthern pots alone are almost always sufficient.

"But what most concerns the subject of this survey [the navigation of the Euphrates], is the existence of a parapet wall or stone rampart in the river, just above the several aqueducts; in general there is one of the former to each of the latter, and almost invariably between two mills on opposite banks, a wall crosses the stream from side to side, with the exception of a passage left in the centre for boats to pass up and down. The object of these sub-aqueous walls (mistaken by Alexander the Great for means of defence against his irresistible legions) would appear to be entirely with a view to raise the water sufficiently at low seasons to give it impetus, as well as a more abundant supply to the wheels; and their effect at times is to create a fall in every part of the width, save the opening left for commerce, through which the water rushes with a moderately irregular surface: these dams were probably from four to eight feet high originally; and they are now frequently a bank of stones, disturbing the evenness of the current, but always affording a sufficient passage for large boats at low seasons; and ceasing to be very perceptible, except by the broken surface, after the water is swollen.

"Ten miles below the town of Hit, the last of these irrigating mills and artificial barriers is passed, and a few miles lower the double range of hills is nearly lost, and the country becomes comparatively flat."

The water-wheel used in China unites with the simplicity of all Chinese mechanism great ingenuity of construction and adaptation. The change of position of the buckets, if they may be so called, produced by the revolution of the wheel, without any motion of the buckets themselves, well deserves to be noticed.

Two hard wooden posts or uprights are firmly fixed in the bed of the river in a line perpendicular to its banks; these posts support the pivots of an axis of about ten feet long: this is the axis of a large wheel, consisting of two rims of unequal diameter, the rim which is nearest to the river's bank being fifteen inches less than the other; but both rims dip into the stream, and both rise above the trough or spout which receives the water and conveys it to the land to be irrigated.

The only materials employed in the construction of this water-wheel, except the axle and the two posts on which it rests, are afforded by the bamboo. The rims, the spokes, the ladle-boards or floats, the tubes or buckets, are made of entire lengths, or large pieces, or thin slices, or single joints, of bamboo; neither nails, pins, nor screws, nor any kind of metal, are used; the parts are firmly bound together by cordage of split bamboo or cane.

The wheel is thus framed: sixteen or eighteen spokes are inserted obliquely into the axis near each extremity, and cross each other at about two-thirds of their length; the wheel is strengthened where the arms cross by a concentric circle, and the ends of the arms or spokes are secured to the two rims. The spokes proceeding from the inner end of the axis next to the river's bank support the larger or outer rim, and those from the outer or inner end of the axis support the smaller rim, which is next to the land.

Between the rims and the crossing of the spokes is a triangular space, which is woven with a kind of close basket-work to serve as ladle-boards or floats; these, successively entering the stream, are impelled by the current, and thus motion is imparted to the wheel.

The buckets are tubes of bamboo, closed at one end and attached to the two rims of the wheel; and so fixed that when they are on a level with the axle, they have an inclination of twenty-five degrees to the horizon and to the wheel's axis.

The closed end of the tube is of course the lowest, and it is fastened to the outer and larger rim; the open end is attached to the smaller rim nearest the land. By this position, the buckets which dip into the stream as the wheel revolves, fill with water and rise with their mouths uppermost; their inclination gradually alters as they rise, but not so much as to let their contents flow out until they reach the top, when they pour the water into a wide trough, from which it is distributed to the plantations.

These wheels are from twenty to forty feet in diameter, according to the height of the land on the river's bank and the consequent elevation to which the water must be raised. A wheel of thirty feet carries twenty tubes or buckets, about four feet long and two inches inside diameter, each of them

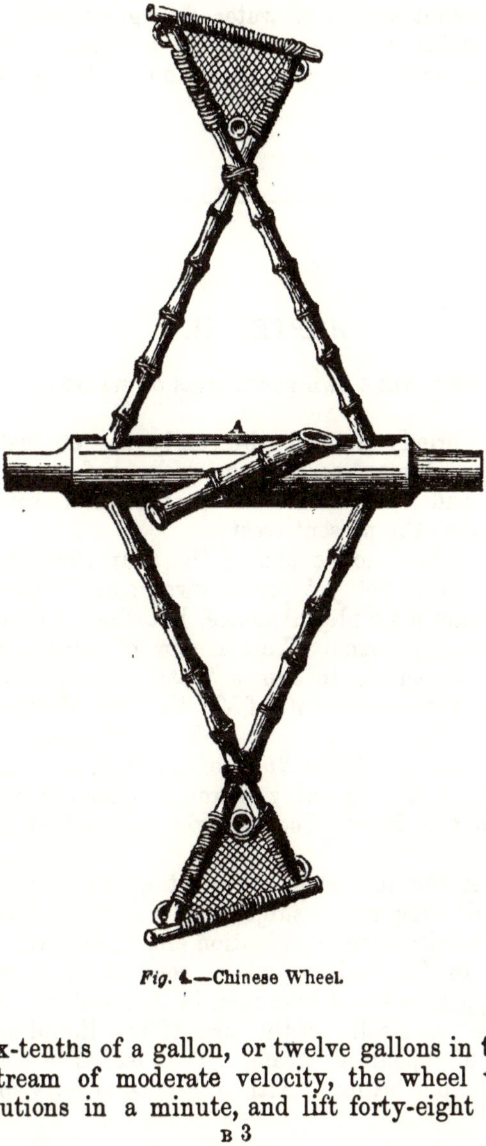

Fig. 4.—Chinese Wheel.

holding six-tenths of a gallon, or twelve gallons in the whole. With a stream of moderate velocity, the wheel will make four revolutions in a minute, and lift forty-eight gallons of water, or 2,880 gallons in an hour, or more than 69,000 gallons per day.

The position of the bucket as it rises to the level of the axle is shown at A Fig. 4.

Thus, at a very trifling expense, a machine may be constructed which, without labour or attendance, will furnish a large and constant supply of water for agricultural purposes at a considerable elevation. There are many places in England where such means of irrigation might be used with advantage.

CHAPTER II.

THE NATURE AND PROPERTIES OF WATER.

Having thus briefly noticed the early use of water as a motive power, it may perhaps be well, before proceeding further, to consider the nature and properties of water itself, so far as concerns the present treatise.

It was long held to be one of the four elements, and a hundred years have not yet passed away since it was proved that water is not a simple substance, but that it is composed of oxygen and hydrogen. The honour of announcing this fact appears to belong to James Watt; the researches of Dr. Black, and the experiments of Mr. Warltire, Mr. Cavendish, and Dr. Priestly, led him to this conclusion, which is stated in a letter dated the 26th of April, 1783, addressed by Mr. Watt to Dr. Priestley, through whom it was communicated to the Royal Society, who printed it in their Transactions.

The idea of the four elements must now be treated only as a beautiful allegory, a subject on which sculptors and artists may exercise their imagination and their skill.

According to Dr. Dalton, water consists of 8 parts by weight of oxygen and 1 of hydrogen in 9 of water. It exists in four states: as a solid in the form of ice, liquid as water, in the state of vapour as steam, and it exists also in combination with other bodies. Although it is, strictly speaking, in the liquid form alone that we have now to consider it, yet it may be noticed, as important to the present subject, that water expands in bulk and decreases in density from a temperature of 39 degrees of Fahrenheit's thermometer up to 212, when it boils and evaporates into steam; that below 39 degrees it again expands and decreases in density down to 32 degrees, when it crystallises into ice.

It is seldom that this difference of density and volume is of much consequence in mechanical practice, yet it is most important as it respects the great operations of nature; the expansion of water in assuming the form of ice should never be forgotten by those who have to construct hydraulic works; its bulk is then increased in the proportion of 9 to 8, and the force with which it expands is so great that scarcely anything will resist it. The strongest pipes and vessels of iron are split, the heaviest weights are lifted, masonry and even rocks are rent asunder, when water having found its way through some small chink or opening, has frozen within the mass; and experiments have rendered it probable that a single cubic inch of water confined and frozen, exerts within the range of expansion a force equal to 13½ tons. It is therefore absolutely necessary that care be taken either to prevent water from lodging or to provide means for allowing it to expand when it freezes. By thus becoming lighter than water, ice floats upon the surface, and by exposure to the sun's rays is quickly melted.

The earth, however, is a good non-conductor of heat, and in the temperate climate of England, the frost seldom reaches more than 10 or 12 inches below the surface of the ground in the coldest winter, so that water-pipes laid at the usual depth of two feet underground are sufficiently protected from freezing. The combination of water with other bodies, although a chemical operation, is most valuable to those whose business it is to build dams and sluices. If water be thrown on quick lime, nothing but a red heat will again separate them; plaster of Paris in the state of powder becomes converted into a compact mass when mixed with water. Barrow lime and other limes of similar character from the lias beds, and Roman cement, become solid and remain hard under water.

A cubic foot of rain-water at a mean temperature, when the barometer stands at 29·5 inches, weighs 1000 ounces avoirdupois, or 62½ lb.; consequently water is generally taken as the standard in tables of specific gravity, in which the weight of equal bulks of other substances are compared with that of water in ounces to the cubic foot.

A cubic foot contains 1728 cubic inches, and the imperial standard gallon 277·274 inches; a pound avoirdupois weight of water contains 27·64 cubic inches; so that an imperial gallon may be taken to weigh 10 lb.; and a cubic foot to contain 6¼ gallons.

A pipe of one inch in diameter, and one yard in length, contains 28·26 cubic inches, or very nearly a pound of water. Hence the following practical rule is generally used by millwrights to find the quantity of water in a pipe of any given diameter:—

Square the diameter of the pipe in inches, and you have the weight of water in pounds per yard of the pipe's length; shift the decimal point one place to the left and you have the quantity of water per yard in gallons.

Thus if a pipe be 12 inches in diameter, 144 will express the number of pounds weight of water it contains in every yard of its length, and 14·4 the quantity in gallons, with sufficient exactness for any practical purpose in mill-work.

The force required to compress water, even to render the diminution of its bulk appreciable, is so great, that practically it may be regarded as incompressible, and consequently non-elastic. Hence it is that when a strong iron vessel, as the cylinder of a hydraulic press, is burst by overloading it, no explosion takes place; the breaking of the metal at once relieves it from a force so limited in its range of expansion, that the fragments do not fly asunder, but fall as if broken by mere weight.

The author witnessed the compression of water by the apparatus of Mr. Perkins; and Professors Colladon and Sturm, in a series of experiments carefully made on water completely deprived of air, reduced its bulk $\frac{1}{890}$ with a pressure of 24 atmospheres.

This resistance to compression renders water so useful a medium for transmitting and multiplying power by means of Bramah's press, and thus by the hydraulic ram, the impulse of a fall of water a few feet in height through an enclosed pipe, when suddenly checked, strikes as it were a blow which will force a portion of the stream much higher than the level of the dam from which it flows.

Water in falling is subject to the same laws of gravitation as other heavy bodies, but in the operation of those laws there is this difference; namely, that a detached body, as a stone, a mass of metal, a cannon-ball for instance, falls freely and without other retardation or resistance than that of the air, or without any retardation, if it fall *in vacuo;* but water to fall must flow also, consequently in speaking of the fall of water, we mean a continuous stream, or jet or sheet of water descending from a higher to a lower level, all parts of which are in motion at the same time, and in constant connection with the succeeding column and adjacent body of water, to which also motion is communicated, and consequently force is lost.

If a ball of lead be dropped from a high tower, it falls (as nearly as may be) 16 feet 1 inch in one second of time; at the end of that time it has acquired a velocity of 32 feet 2 inches; it falls in the next second, 48 feet 3 inches, and in two seconds of time it has fallen 64 feet 4 inches. So that if the time or seconds be represented thus, the time 1″ 2″ 3″ 4″ 5″; the spaces fallen in each second will be as 1, 3, 5, 7, 9; the spaces fallen through in the whole time as 1, 4, 9, 16, 25. This preportion holds good for any time or space through which a body falls *in vacuo;* but the air's resistance in the present case does not materially interfere, and may be disregarded within the limits of a fall of water applied to mechanical purposes.

Thus the velocities are as the times of descent, and the spaces fallen through are as the squares of the times; consequently, the velocities are as the square roots of the heights or spaces through which a heavy body falls; therefore, as gravitation produces the velocity of 2 in descending through the space 1, the height in feet through which the body falls, being multiplied by 64·3, will give the square of its velocity, in feet, per second. So if the height fallen be 1 foot, the square root of 64·3, or 8·02 (very nearly), is the velocity with which a heavy body falls through that space.

The difference between a flowing and a falling body is evident, although both are governed by the laws of gravitation, and accordingly we find that the motion of fluids has attracted the attention of mathematicians and philosophers, from a very early period, in Italy, France, Germany, and England, up to the present day; and many experiments have been made from time to time, to ascertain exactly the value of this difference. The late Mr. Rennie, being much engaged in the improvement of rivers, the drainage of large tracts of land, and the application of water to turn mills, made many elaborate experiments for this purpose; his researches have been ably continued by his sons, and Mr. George Rennie has contributed to the Philosophical Transactions some valuable papers on the subject: subsequently, an extensive series of experiments was made by MM. Poncelet and Lesbros, by order of the French Government.

The value and importance of such investigations will be apparent, when it is considered that the first points to be determined, in all great hydraulic works, are the quantity of water available for the object in view, and the height from which it falls.

These constitute the power to be employed; and when these are known, the application of the disposable force to produce the greatest effect is next to be considered.

The quantity which a stream will yield in a given time, is generally ascertained by placing a dam or barrier across it, commonly of timber. In this there is either a sliding sluice, through which the water is discharged, so that the size of the rectangular orifice can be regulated at pleasure; or else a rectangular notch is cut in the edge of the dam, at the surface of the water, so that the section of the passing stream may be measured as it flows through the notch.

If the water flow through an opening in the dam, so adjusted by a sluice that the discharge runs constantly through it, while the dam is maintained either full or at an uniform level above the opening; and if the laws of gravitation, without allowance or correction, governed the flow, it would only be requisite to measure the area of the opening, and to calculate the velocity of the water as that of a heavy body falling from the surface of the dam or reservoir through the centre of the orifice, and the area multiplied by the velocity would give the quantity of water passing through it.

In like manner, if no correction were required, to find the quantity of water flowing through a rectangular notch, cut in the upper plank of the barrier, it would only be requisite to measure the height from the surface of the reservoir to the bottom of the notch, in order to find the velocity as before; and then to take two-thirds of the quantity which would flow at that velocity through the whole area of the notch, the proportion of two to three, that is the area of the parabola described by the water, being two-thirds of its circumscribing rectangle.

Such, indeed, were the rules given by some of the best

authorities; but the evidence of the senses cast a doubt on their truth, and experiments showed that the actual quantity of effluent water was less than the theoretic quantity.

Sir Isaac Newton found that the velocity of water issuing from a round hole in the bottom of a vessel four feet high, the hole being an inch in diameter, and made through a thin plate of metal, was less than it ought to be in the proportion of 1 to the square root of 2; for he observed the vein of the effluent water, and found it to contract, and grow narrower, to the distance of about a diameter of the hole below the bottom of the vessel, so that the diameter of the vein at this place was less than the diameter of the hole, in the proportion of 21 to 25, and consequently, the area of the section of the vein thus contracted, to be less than the area of the hole, in the proportion of 441 to 625; that is, as 1 to the square root of 2, or as ·7056 to 1.

But as all experiments of this kind, made by mathematicians, were of necessity conducted on a small scale, the results deduced from them were by no means satisfactory to men who had to deal with water in the execution of large works, as it was not probable that the causes of contraction in a vein of water an inch in diameter, would bear the same proportion to the waters of a river discharged through a sluice.

Hence it was, that persons in extensive practice endeavoured to form rules for themselves, based on their own experience; and as the circumstances in which such observations were made varied, so did the rules deduced from them vary also. Mr. Rennie made many experiments and observations to determine the difference between the theoretic discharge (computed by the laws of gravitation) and the actual discharge diminished by friction, lateral retardation, reaction of the adjacent fluid, and other causes of diminished velocity or volume, and consequent quantity. Other engineers did the same. The French Government, considering the question as one of public interest, appointed a commission to determine it; the elaborate experiments on the great scale, made by their engineers, under the direction of MM. Poncelet and Lesbros, brought it within narrow limits; and, in December, 1851, the Institution of Civil Engineers, in London, awarded an honorary premium to Mr. Blackwell, one of their members, for his paper, entitled, "Results of a series of practical experiments on the discharge of water by overfalls."

CHAPTER III.

ON THE MEASUREMENT OF EFFLUENT WATER.

There is, perhaps, no point which has occasioned more dispute and litigation, than the conflicting rights of persons entitled to take water power, in certain proportions, from a common source, where the demand exceeds the supply; and there are, perhaps, few of greater interest at the present time, when the increased size and population of our towns and cities render their call for water imperative.

The conclusions arrived at by Mr. George Rennie, as reported in the Philosophical Transactions of the Royal Society, are:—

1. That the quantities discharged in equal times are as the areas of the orifices.

2. That the quantities discharged in equal times, under different heights, are *nearly* as the square roots of the corresponding heights.

3. That the quantities discharged in equal times, under different heights, are to each other in the compound ratio of the areas of the apertures, and of the square roots of the heights, *nearly*. The heights were measured from the centre of the apertures; and the mean of several experiments showed that the co-efficients, or numbers, expressing the proportion between the theoretic discharge of the water, calculated as a falling body, and the actual discharge, as measured, are as under; all the openings being formed in brass plates $\frac{1}{20}$ of an inch thick.

Round hole, 1 inch diameter, with	4 ft.	head.	0·621
Do. do.	1 ft.	,,	0·645
Triangular hole (equilateral) 1 in. area	4 ft.	,,	0·593
Do. do. do	1 ft.	,,	0·596
Rectangular holes 1 in. sq. and 2 × ½	4 ft.	,,	0·593
Do. do. do. do.	1 ft.	,,	0·616

The mean of all these numbers is ·610, and for the rectangular holes, it is ·600. Hence Mr. Rennie deduces the following formula:—

Let A = the area of the orifice in square feet.
H = the altitude or head of water in feet.
T = the time in seconds.
g = the action of gravity in one second.
Q = the quantity of water discharged in cubic feet.

Suppose the hole to be 2 feet long and 6 inches wide, the area 1 square foot, having the long side parallel with the water's surface, and a head of 4 feet above the centre of the hole. Then will $Q = \cdot 6\ A\ T \sqrt{g\ H}$.

Thus—multiply 64·3, the effect of gravity, by 4 feet, the head of water, for the theoretic velocity in feet per second; and as the area is 1 square foot, the velocity will also be the quantity discharged per second in cubic feet.

	64·3
	4
1	257·2(16·037
1	1
26	157
6	156
3203	12000
3	9609
32067	239300
	224469

15·037
·6

Actual discharge 9·6222 cubic feet per second.
or, 577·332 cubic feet per minute.

The results of the experiments made by MM. Poncelet and Lesbros, by order of the French Government, in which neither labour nor expense was spared, show that with an opening of 20 centimètres or 7·87 inches square, and with a head of 1 mètre 68 centimètres, or 5 feet 6 inches from the upper side of the opening to the surface of still water, the co-efficient is 0·600. When the head or altitude was diminished to four or five times the height of the opening it increased to 0·605, but it again diminished rapidly, as the altitude diminished to 0·593.

With orifices of smaller dimensions; namely, from 10 to 5 centimètres (3·9371 to 1·9685 inches) the same law was observed, the co-efficients being, for an opening of 10 centimètres, 0·611, 0·618, and 0·611; for an opening of 5 centimètres 0·618, 0·631, and 0·623.

It is evident that with small orifices the effect of high heads would be to contract the vein and diminish the discharge, and the nearer the orifice could be brought to the surface, so that it might still be kept running with a full stream, and without causing any depression of surface or eddy, the greater would be the discharge. But with larger apertures, of a mètre in length, by 50 centimètres in depth, or 39·371 × 19·685 inches, equal to 5·38 square feet in area, the co-efficients gradually increased with an increased head or altitude, thus:—

A head of	0·10	mètres =	3·937 inches,	gave a co-efficient of	0·552
„	1·00	„	39·371	„	„	0·591
„	4·00	„	157·484 or 13 ft. 1½ in.	„	„	0·604

All these observations and experiments differ so little, that ·6 of the theoretic quantity may be taken for general practice.

The proportion between the theoretic and the actual discharge from open notches in dams, was found to be ·677 when the notch was 9 inches deep, and ·727 when it was 1 inch deep; and as the discharge from an open notch is equal to ⅔ of that from an orifice of the same size discharging a full stream under the same head, these numbers being multiplied by ·666, give ·450, and ·484 as the factors for finding the quantities of water issuing from notches of those depths in a second of time—for greater depths, and as a general rule, ·400 may be used. Mr. N. Beardmore, in his useful handbook of tables, showing the discharge through sluices, pipes, &c., founds his calculations of the flow of water through open notches on the following formula:—

$$D = 214 \sqrt[2]{H^3}$$

taking D to be the quantity discharged in cubic feet per minute over 1 foot in width of the waste board or sill of the notch, and H to be the true height from the sill of the notch to the surface of the water where it is at rest. The principle of this formula, as in those already noticed, is, that the curve of the water falling over being a parabola, there can be discharged only two-thirds of the water that would pass the full section due to H; the constant number 214 is two-thirds of 321, which he states has been found by frequent trials to represent the factor to be multiplied by the square root of H, the height in feet for giving the mean velocity in feet per minute of water passing over a waste board.

Assuming this to be correct, the velocity, and consequently the quantity of water which would run over every foot in width of a rectangular notch, 1 foot in depth from the water's surface, would be 214 cubic feet per minute.

The square root of 64·3, or 8·02, which is the theoretic velocity for 1 foot fall, multiplied by 0·45, gives for the actual velocity, 3·001 feet per second, and this by 60 gives 216 cubic feet per minute.

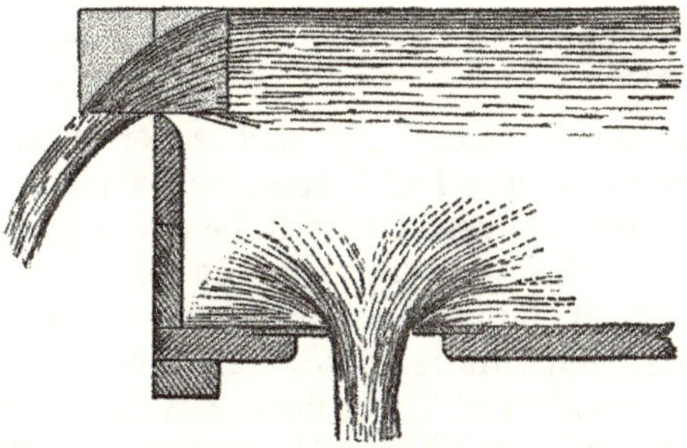

Fig. 5.

This difference is very small compared to what may arise from the form of the notch or aperture; for a plain rectangular notch with square edges cut in a 3-inch plank, will discharge very much less than one having its inner edges next the reservoir, or dam whence the stream issues, bevelled off in the parabolic form of the contracted vein. If the notch or aperture be of small dimensions, a difference of one fourth of the whole quantity may be made, as was proved by Venturi in works on a larger scale; care should be taken to form the wing walls to sluices, and bridges, with parabolic or "trumpet-shaped" approaches, so that the water may enter and pass without other obstruction than the contraction of the overflow, or sluice-way itself may present to it; and when water passes through a parallel channel, canal, or trunk, the sectional area, multiplied by the mean velocity, will show the quantity passing in a given time.

The mean velocity is that of the water's surface added to that at the bottom of the current, and their sum divided by two, or it may be determined sufficiently near for any useful purpose, by ascertaining the surface velocity in inches per second in the middle of the stream; call this velocity s, and the mean velocity will be equal to $s - \sqrt{s - 5}$. Suppose the stream flows at the rate of 36 inches per second, or 180 feet per minute upon the surface, then

s being equal to 36 inches		36
deduct	5	5·5
extract the	5 | 31·(5·5	30·5 in. per second.
square root	5 | 25	60
	105 | 600	12 | 1830·0
	525	152·5 ft. per minute.

If 30·5 the mean velocity in inches per second be multiplied by 5, the product is 152·5, the mean velocity in feet per minute. The bottom velocity will be equal to $s - \sqrt{s} \times 5$ or 36—11=25 inches per second, or 125 feet per minute, very nearly; the sum of the top and bottom velocities so calculated is 305, and half that sum is 152·5. If, therefore, the water-course be 4 feet wide and 2 feet deep, having a sectional area of 8 feet, 152·5 × 8 or 1220 cubic feet will pass through it in one minute.

While writing these pages, the author received a copy of Mr. Blackwell's paper containing the details and results of his numerous and varied experiments.

Mr. Mylne, of the New River, kindly sent a series of experiments, in making which he was assisted by Mr. Murray, then his pupil; and other engineers, engaged in water-works on a large scale, contributed their observations, and the practical rules they had founded upon them.

In principle all these rules agree; their corrections vary less than might be expected; and their expression is generally similar to the formula given by Mr. George Rennie, in his paper read to the Royal Society, which Mr. Blackwell has called *No.* 1, and writes thus:—

$$Q = \sqrt{2\,g\,H} \times l\,H \times m.$$

in which

Q is the discharge in cubic feet per second.
$2\,g$ — 64·3 the effect of gravity (Mr. Rennie's g).
H the head in feet (l H the section of the stream).
l width of the overfall in feet.
m the co-efficient of correction.

Or they resemble the following formula, which he calls No. 2, and writes thus:—

$$Q = H^{\frac{3}{2}} \times l \times k$$

in which

Q is the discharge in cubic feet per second.
l the width of overfall in feet.
H the head in inches.
k the co-efficient of correction.

The head is the height of still water above the sill of the overfall.

Mr. Mylne's rule runs thus:—Extract the square root of *two-thirds* of the head or depth, measured in inches, from the sill of the overfall to the surface of the still water, and multiply it by sixteen for the velocity in inches per second; multiply this velocity by the section of the stream, also in inches, and divide the product by 1728, the number of inches in a cubic foot, for the quantity in cubic feet discharged per second. A practical allowance for retardation is here made, by taking two-thirds of the head instead of the whole.

With respect to the formulæ, No. 1 and No. 2, which are in general use, the only difference among engineers is in the numbers they take as co-efficients of correction. The number commonly taken by English engineers is 5·1 or 5·15 per foot per minute.

Thus, if the depth of overfall be 12 inches, the cube is 1728, and the square root of this, 41·58 · 5·15, gives the discharge through an open notch a foot square, as 214·1876 cubic feet per minute, the same as shown in Mr. Beardmore's tables.

Mr. Mylne's rule gives 226·2 cubic feet. Some engineers use 5·35 and 5·4, which last number gives 224·532 cubic feet per minute. Mr. Blackwell's experiment, like those of the French engineers, MM. Poncelet and Lesbros, show that the co-efficients vary with the depth, the width, and the form of the notch or overfall, and that all the numbers used are, strictly speaking, applicable only to the peculiar circumstances of each particular case, although, within certain limits, they may approximate sufficiently near to the truth for practical purposes.

These experiments, 243 in number, were made by Mr.

Blackwell on the Kennet and Avon Canal, through overfalls varying in width and form.

1. With the water falling over the edge of a thin iron plate, through notches 3 feet and 10 feet long, of varying depths.

2. Over a plank, 2 inches thick, through notches, 3 feet, 6 feet, and 10 feet long.

3. Over the same plank, 2 inches thick, with wings converging towards the notch at an angle of 64 degrees.

4. The notch, 3 feet wide, with the sill or crest, 3 feet broad, sloping 1 in 12, fixed on to the outer edge of the plank, so as to form an uninterrupted continuation of it, like the crest of a weir.

5. The same, with the crest or sill sloping 1 in 18.

6. The overfall, with notch 10 feet long, the crest or sill 3 feet broad, laid level.

Similar experiments (about twenty in number) were made at Chew Magna, under the direction of Mr. James Simpson, which are tabulated in Mr. Blackwell's paper with his own; and this table gives perhaps the best arranged view of the subject which has yet appeared. The reader who desires to know the details of these experiments is referred to Mr. Blackwell's communication of May 6, 1851, in the Proceedings of the Institution of Civil Engineers—a paper which well deserves perusal and study.

TABLE *showing the* VARIATION *of the* CO-EFFICIENTS *for the different* OVERFALLS. (*m* and *k* mean Co-efficients.)

Overfalls.		*m*	*k*
Thin plate,	3 feet long	·421	·080
"	10 "	·445	·086
Plank, 2 inches wide,	3 "	·380	·073
" "	6 "	·377	·072
" "	10 "	·371	·072
" (with wings)	10 "	·459	·090
Bar, 2 inches wide, Chew Magna,	10 "	·480	...
Crest, 3 feet wide, sloped 1 in 12	3 "	·338	·065
" " 1 " 18	3 "	·339	·065
	10 "	·337	·065
level	3 "	·311	·060
" " "	6 "	·322	·061
" " ,	10 "	·314	·061

It will be observed that the means of these numbers for

simple overfalls or notches, excluding those given by the broad crests, made to resemble the top of a stone weir or dam, are for *m* ·419, and for *k* ·079; which last number being multiplied by 60, to make it the factor for feet per minute, gives 4·74. So that the first is slightly over Mr. George Rennie's ·400, and the second somewhat under the number 5·1 or 5·15 generally used. This is accounted for by so many experiments with low heads, as will be seen on referring to the next table, arranged by Mr. Blackwell.

TABLE *showing the* VARIATION *of the* CO-EFFICIENTS *for the different* HEADS OF WATER. (*m* and *k* mean Co-efficients.)

No. of trials.	Description of Overfalls.	Head in inches.	*m*	*k*
6	Thin plate, 3 feet long . .	1 to 3	·440	·085
	" " " . . .	3 to 6	·402	·078
11	Thin plate, 10 feet long . .	1 to 3	·501	·096
	" " " . . .	3 to 6	·435	·086
	" " " . . .	6 to 9	·370	·072
23	Plank 2 inches thick, with notch 3 feet long	1 to 3	·342	·066
		3 to 6	·384	·074
	" " " . . .	6 to 10	·406	·077
56	Plank 2 inches thick, with notch 6 feet long	1 to 3	·359	·069
		3 to 6	·396	·077
	" " " . . .	6 to 9	·392	·074
	" " " . . .	9 to 14	·358	·069
40	Plank 2 inches thick, with notch 10 feet long	1 to 3	·346	·068
		3 to 6	·397	·076
	" " " . . .	6 to 7	·374	·072
	" " " . . .	9 to 12	·336	·062
4	Plank 2 inches thick, notch 10 feet, with wings	1 to 2	·476	·092
		4 to 5	·442	·087
7	Overfall with crest	1 to 3	·342	·066
	3 feet wide, sloping 1 in 12 . .	3 to 6	·328	·063
	3 feet long, like a weir . . .	6 to 9	·311	·060
9	Overfall with crest	1 to 3	·362	·070
	3 feet wide, sloping 1 in 18 . .	3 to 6	·345	·066
	3 feet long, like a weir . .	6 to 9	·332	·064
6	Ditto. Sloping 1 in 18 . . .	1 to 4	·328	·063
	3 feet wide × 10 feet long . .	4 to 8	·350	·068

No. of trials.	Description of Overfalls.	Head in inches.	m	k
14	Overfall, with level crest . .	1 to 3	·305	·059
	3 feet wide × 6 feet long . .	3 to 6	·311	·060
	" " " . . .	6 to 9	·318	·061
15	Overfall 6 feet long, with level	3 to 7	·330	·062
	crest 3 feet broad . . .	7 to 12	·310	·060
12	Overfall with level crest, 10 feet	1 to 5	·306	·059
	long, 3 feet broad . . .	5 to 8	·327	·063
	" " "	8 to 10	·313	·061
61	At Chew Magna, overfall bar, 10	1 to 3	.437	
	feet long, 2 inches thick .	3 to 6	·499	
	" " " . . .	6 to 9	·505	

It appears from the detailed experiments, that when the overfall is a thin plate it discharges a greater proportionate quantity when the stream is only 1 inch deep than with greater depths, the vein contracting with the increased head, and consequent force of the stream. When planks 2 inches thick form the overfall, the flow of water being more retarded, a greater head is requisite to overcome the resistance; and the maximum discharge is given by a head of 7 inches; and it does not afterwards increase in proportion to the depth, but the co-efficient lessens as the depth becomes greater. When the length of the overflow plank is 10 feet, the co-efficient is greatest with a depth of 5 inches, namely ·406; and when wing-boards are added, causing the stream to converge towards the overfall, at an angle of 64°, the co-efficient is greater even when the head is less, and shows the great utility of proper wing-walls on bridges and sluices. When the overfall has a broad crest like a weir, the co-efficient, as might be expected, decreases, and is least when the crest is level.

The author has been anxious to avail himself of the many observations and experiments with which he has been favoured on this part of his subject, respecting which so much difference of opinion has existed, and to bring the whole, which were scattered in different places, and in several hands, into one general view, even at the risk of being somewhat tedious to his readers. The public is much indebted to Mr. Blackwell for having laid these numerous trials before the Institution of Civil Engineers; and it ought to be more generally known that such valuable papers are not merely given to members, but may be purchased from the booksellers' by all who wish to possess them.*

CHAPTER IV.

THE SOURCES AND SUPPLY OF WATER, ITS DISTRIBUTION AND USE AS A MOTIVE POWER.

Our limits will not admit of our entering into those details which would be requisite in order to lay a clear account of a few of the causes which are known to have an influence in regulating the distribution of rain. We must therefore confine ourselves to a very brief general outline. There are many good reasons for believing that the great reservoir which supplies the moisture brought to the British coast by westerly winds is the trade-wind zone of the Southern Ocean. The moisture is uniformly distributed through large masses of the moving air, but it is precipitated in unequal quantities over areas of limited extent. This inequality is influenced by the form of the surface of the land, and by its varying altitude. The moisture-bearing south-west wind generally flows along the surface of the Atlantic, and its degree of humidity is greatest where it approaches that surface. When the current reaches land it dashes against it where it is abruptly elevated, or is deflected upwards where the rise is gradual. Hence in a hilly district the distribution of rain is more irregular than in a flat one, and the amount of rain will be greater in those valleys or on those slopes in proportion as they directly face the prevalent rain-bringing wind. We will select a couple of instances, out of hundreds which we might quote, to show how close is the connection between the surface configuration of a country and the amount of rainfall.

If a line be drawn in an east and west direction from Mahabuleshwar on the summit of the Ghâts in India to Phultun, a distance of 40 miles, we shall find at the com-

* The reader desirous of more detailed and more precise information on the subject referred to in this chapter, should consult "The Hydraulic Tables" of Mr. Neville, C.E., the Translation of De Voisin's "Hydraulics," by Professor S. Downing, C.E., and the "Traité d'Hydraulique" of General Morin.

mencement of the line a rainfall of 240 inches in the year, the elevation of the station above the sea being 4,500 feet; of 180 inches at Sindola, 1 mile off, and with an altitude of about 4,600 feet; of 50 inches at Paunchgunnee, 11 miles farther, altitude 4,000 feet; 25 inches at Wye, 4 miles farther still, altitude about 2,300; and at Phultun, having an elevation about equal to Wye, 7 or 8 inches. It is probable that the quantity assigned to the last place is not quite correct. The first two cases are remarkable. Between Mahabuleshwar and Sindola, an intervening eminence, Mount Charlotte, rises 100 feet higher than Sindola. A still better example may be quoted, in which all the localities are situated in the Uttray Mullay range.

In 1849 the fall at a station at the base of the range and

500	feet above the sea level was		99	inches.
2,200	,,	Attagherry	170	,,
4,500		Uttray Mullay	240	,,
6,200	,,	Agusta Peak	194	,,

This and many other examples which have been collected by Colonel Sykes tend to prove that in the Ghâts the amount of rainfall increases with the altitude up to about 4,500 feet, after which there is a decrease with the height. This limitation appears to be due to the moisture-bearing current being generally less than 4,500 feet in depth; above it is a stratum of comparatively dry air.

Our own country affords similar examples; the most remarkable of these is the lake district of Cumberland and Westmoreland. It has long been noted for the remarkable abundance of its rains, and the great local differences in the quantity. Its hills are lofty, and are situated near the coast. The land is broken up into a compact group of ridges and valleys, which radiate from a central point. The prevailing wind which accompanies rain in these hills is the south-west, and this is because there is no other wind which contains so large a quantity of aqueous vapour. This wind, then, charged with moisture gathered in its course across the Atlantic, strikes against the hills of the lake district, and by their peculiar shape is compelled to rush up the valleys. As it ascends the valley, it becomes cooler, and a greater quantity of vapour is condensed. It follows, therefore, supposing the theory above suggested to be true, that the higher a place is situated in the same valley the greater should be the rainfall. That this is so, is shown by the following examples which we have selected as being the most striking, both for the great quantities of rain recorded, and the extraordinary differences exhibited. Loweswater, Buttermere, and Gatesgarth are all in the same line of valley, and within a space of 4 miles. In 1848 the rainfall at Loweswater was 76 inches, at Buttermere 98 inches, and at Gatesgarth 133½ inches. The same law of increase with height is evident at places situated a few yards instead of miles asunder. For instance, a gauge was placed at Wastdale Head, within 200 yards of another one higher up the valley. The gauge nearest the summit of the hill received a slight excess of rain over the other every month, and at the end of the year the difference between them amounted to 3·53 inches. So close is this connection between the form of land and the distribution of rain, that it is almost possible to name the amount of rainfall of a locality on knowing its geographical position, the prevailing wind, and the precise form of the surrounding country. In the Cumberland hills, for example, having all the facts of situation, wind, &c., before us, and supposing the valleys to be of the same height, and to have the same slope, it might be predicted that the valley which most directly faces the south-west wind would have the greatest rainfall. Such is the case with Gatesgarth; it is at the head of a valley which looks towards the south-west. A little to the east of this is Wastdale Head, and this comes next in quantity and difference. For a long time Seathwaite was regarded as the wettest place in Great Britain, but the fall in the localities just mentioned is now known to be greater than at Seathwaite. So on the other hand, when we know the particulars of the rainfall in a district we can, to a certain extent, judge of its probable configuration. In Cumberland as in India the amount of rainfall increases with the altitude up to a certain point (which is about 2,000 feet in Cumberland), and decreases above it. Since large slopes, such as mountain valleys, influence the reading of a gauge, so, doubtless, small slopes, such as the sides of small hills, have their effects. A gauge should, therefore, be placed on a level, because if situated on a slope it will receive less rain than if at the same height on the *top* of a hill.*

Unless the position of the gauge and of the locality be

* Mr. Miller, in his paper on "The Meteorology of the Lake District" (Phil. Trans., 1849), shows that at Seathwaite, near Derwentwater, 422 ft. above the sea, 143·96 inches of rain fell in 1850; and at a station on Sprinkling Fell, 189·49 inches. In 1861 Seathwaite had 182·58 inches, and in 1863, 173·84.

known, it is clear no fair comparison can be made between one place and another, or fair inferences drawn as to the distribution of rain. For the last five or six years elaborate tables of the rainfall in the British isles have been compiled by Mr. G. J. Symons, from observations made at some hundreds of localities, detailing the amount of rainfall for every month in the year. In these tables the height of the rain-gauge above the ground and above the sea level is given. Some of the observers have placed their gauges in the ground, and others at heights varying from this point up to 70 or 80 feet. Under these circumstances a person consulting these tables, and neglecting differences in the position of the gauge, and the physical peculiarities of the district, might arrive at wrong conclusions. As long as the effects produced by hills and by the elevation of the gauge above the ground are neglected, a comparison of the rainfall in two places must lead to fallacious results.

The influence of the form of the ground should always be considered apart from the influence of the position or construction of the rain-gauge. The neglect of this last consideration has led to many errors of observation. Thus, when the wind rushes against a house or any other obstacle, the air nearest the house becomes compressed, and flows over and by the sides of the house at a speed greater than that due to the force of the wind alone. The rain-drops falling through such wind are bent out of the perpendicular in proportion to its force. If then two drops fall through air having a given velocity, they will be equally diverted, and will remain parallel to the end of their course. But if two such drops fall in air of different velocities, they will approach nearer to or recede farther from one another, and so increase or diminish the amount of rainfall in the intermediate space. A gauge on the top of a honse will receive less rain than a gauge situated in an open space such as a field. Rain-gauges on the tops or near the sides of houses are therefore comparatively useless; they ought to be placed as near the surface of the ground as possible. It is probable that this may account for less rain falling on a building than at its base; a result which has led to the inference that in an open area a corresponding difference would be found between the rainfall at the surface of the ground and at a height of 120 feet. The inference is, in all probability, an incorrect one. It should always be borne in mind that in registering two things are essential, an accurate instrument and a suitable position.

The average quantity of rain falling in England over the whole surface of the country may be taken at about three feet in depth; and the quantity carried off by evaporation, absorption, and vegetation, has been found to amount to about two-thirds of the whole rain, snow, and hail.

In Ireland the case is nearly the same; for, although the air is moist with mists and dew, which constantly renew and refresh the verdure of the "Green Island," there is but little difference in the average quantity of rain. Sir Robert Kane, who has carefully investigated this subject, shows that it may be taken at 36 inches, and assumes that 12 inches of water finally arrive at the sea; which water may, in its course, become available for the purposes of industry, with a force proportionate to the height through which it falls as it flows onward.

Since Sir R. Kane's investigations much has been done in the way of meteorological observations in Ireland, but without in any way affecting the general accuracy of his conclusion. The fullest information on the subject of the rainfall in Ireland may be found in the works quoted below.* According to these authorities, the average rainfall for Ireland is about 34½ inches.

Providence has furnished mechanical power; it is for man to make it available. The sister kingdom is abundantly supplied with the means; and, although some persons may not estimate the amount of water power so highly as Sir Robert Kane, yet he clearly shows that an enormous quantity is constantly running to waste; and his map, which exhibits, at one view, the principal streams and the elevation of the land in Ireland, renders it evident how much may be done by applying this to useful purposes. He thus sums up his calculations:—

"The average elevation of the surface of the country being 387 feet, the water which flows in our rivers to the sea has an average fall of 129 yards; and now, finally, we may calculate the total water power of Ireland. We have for the total quantity of rain falling in a year 100,712,031,640 cubic yards; of this one-third flows into the sea; that is, 33,237,343,880 cubic yards; or for each day of twenty-

* "Report of the Committee of the Royal Irish Academy on the Meteorology of Ireland," by Dr. Lloyd, in the Transactions of Royal Irish Academy, vol. xxii. part 5, 1855; and "On the Fall of Rain in the British Isles during the Years 1862 and 1863," by G. J. Symons. Brit. Assoc. Report for 1864.

four hours, 91,061,216 cubic yards, weighing 68,467,106 tons.

"This weight falls from 129 yards; and as 884 tons, falling 24 feet in twenty-four hours, is a horse power, the final result is, that in average we possess, distributed over the surface of Ireland, a water power capable of acting night and day, without interruption, from the beginning to the end of the year, and estimated at the force of 3,227 horses' power per foot of fall; or, for the entire average fall of 387 feet, amounting to 1,248,849 horses' power."

Of course, much of this enormous force exists where circumstances prevent its becoming useful; and much passes away in navigable rivers, where it cannot be applied with advantage; but there still remains an immense amount of mechanical power, available for the purposes of industry, and at present unemployed.

Sir Robert calculates the water power of the principal rivers and their tributaries in a very clear and interesting manner. But it is not from the rivers alone that the great mechanical power of water is to be derived. Modern practice has shown that, by throwing up an embankment across a mountain pass, or in the gorge of a valley, which expands itself among the hills, large reservoirs—or rather artificial lakes—may be formed, to receive the water which rolls down the declivities from the bursting rain cloud, and to store it up for future benefit.

Two of the greatest works of this kind are, the embankment of Lough Island Reavy, in Ireland, by Mr. Fairbairn, and the Shaw's Water-works at Greenock, in Scotland, by Mr. Thom.

The mill-owners on the Upper Bann were subject to great disadvantages. They were flooded in winter, and in summer the drought stopped their works. In order to economise the winter floods, which were destructive rather than useful, they consulted Mr. Fairbairn, and he thus reported:—

"Lough Island Reavy, which is the best situated reservoir, is a natural lake, bounded north and south by land of considerable elevation. It has good feeders, which, with the overplus waters of the River Muddock, would give ample supplies, and fill the reservoir once or twice in the year. The present area of the Lough is 92½ statute acres. On this is to be raised 35 feet of water, which may be drawn down to a depth of 40 feet under that height. The area thus enlarged will be 253 acres, equal to 140 acres 35 feet deep; and 113 acres 15 feet deep; making a total of 287,278,200 cubic feet of water."

This reservoir has been completed and in operation for some years.

Reckoning the interest of the money expended upon the works, the cost of maintenance, and the expenses of superintendence and distribution, the annual charge amounts to £700; and it is calculated that 33 tons of water are supplied for one shilling; and as 12 cubic feet per second falling one foot is a working horse power, the value of such an improvement may be estimated.

12 cubic feet at $62\frac{1}{2}$ lb. = 750 lb. × 60 sec. =

45,000 lb. falling one foot per minute;

33,000 lb. raised one foot high per minute, being an effective horse power.

12,000 lb. remaining are allowed for the difference between power and effect.

The Shaw's Water-works are even more interesting. The want of water had been long and grievously felt in the town of Greenock, where the people had not a sufficient supply even for domestic use, and it was generally believed that water sufficient even for sanitary purposes could not be obtained in that locality, until a survey was made by Mr. Thom, who reported, that not only might an ample supply be provided for the use of the inhabitants, but that a large surplus would be applicable as mechanical power and for manufacturing purposes, "to an extent at least equal to all the machinery then impelled by steam-power in and about Glasgow."

The report is dated June 22nd, 1824, and the copy from which the author extracts this information was given to his valued friend, the late James Smith, of Deanston, in August, 1840, by Mr. Thom himself, who has in his own handwriting noted the few differences between the original plan and the actual execution of the works.

In consequence of this report, and under the auspices of Sir Michael Shaw Stewart, a joint-stock company was formed for carrying the plan into effect with a capital of £31,000, and an Act of Parliament was obtained.

In April, 1827, the principal portions of the works appear to have been completed, namely, the great reservoir containing 284,678,500 cubic feet of water, and covering about 295 statute acres of land. The compensation reservoir, containing 14,465,900 cubic feet, and covering about 40 acres, and an auxiliary reservoir (marked No. 3 upon the plans) holding 4,652,800 cubic feet, with an area of ten acres.

Five other smaller reservoirs were made, and extended to hold more than six millions of cubic feet, so that the aggregate quantity stored at once would be 310,000,000 cubic feet of water. The embankment of the great reservoir is 60 feet high, and that of the compensating reservoir 23 feet.

The main water-course, somewhat more than six miles in length, was at the same time completed, and also its branch leading to the eastern line of mills, there being two lines, east and west, and on the 16th of April, 1827, the first mill received its supply of water at the rate of twelve hundred cubic feet per minute; at this rate both lines of mills are supplied. The available annual quantity of water to replenish the reservoirs is more than seven hundred millions of cubic feet.

The distinguishing characteristics of "this scheme" as Mr. Thom calls it, are the following:—Instead of erecting mills and factories on natural waterfalls, on the banks of rivers, in remote and almost inaccessible places,—where immense capital must in the first instance be expended in forming roads and houses for the work-people, as well as a heavy and perpetual charge for carriage to and from the seat of trade,—the water is carried by a conduit from the reservoirs, to a populous seaport town, with a redundant unemployed population, where roads, harbours, piers, and everything requisite for the most extensive manufacture are already formed. Besides, by thus forming artificial waterfalls on advantageous grounds, every inch of fall from the reservoir to the sea is rendered available, whereas by the former mode only a very small part of the fall could in general be employed. In the present case a fall of 512 feet has been made available, of which not more than 20 were formerly occupied or thought capable of being usefully employed. But besides the advantage thus gained by increasing the fall, a still greater advantage is gained from the greatly increased and perfectly uniform supply of water, by the adaptation of the various self-acting sluices, and other simple and effective means of regulation.

The reservoirs are large enough to contain a full supply for six months, so that the surplus of a wet year may be stored up, to provide against a dry season; thus turning to account all the water caught upon the surrounding hills, the greater part of which before ran wastefully to the sea.

The practical effect of the Shaw's Water-works has been to place within the suburbs of a populous town a series of mills and factories driven by power equal to that of thirty-three steam-engines of fifty horses each.

Such power being applied to manufacturing purposes, each factory will pay in wages to the work-people about £10,000 a year, making a total annual amount distributed in wages only of more than £300,000, and employing upwards of 7,000 persons, besides giving an ample supply of water for the use of the town.

A greater addition to the wealth of a small community, in so short a time, and by such simple means, can scarcely be imagined.

CHAPTER V.

RAIN-GAUGES AND RESERVOIRS.—EMBANKMENTS AND THEIR CONSTRUCTION.

The recognised importance of noting and registering the fall of rain, and of observing other meteorological changes during the year, has led to the establishment of such registers in several towns where scientific institutions or libraries have been formed; but as there are many localities where no observations are made and registered, and as the fall of rain is much affected by local circumstances, many engineers have found it necessary, especially in remote and mountainous parts of the country, to institute a series of observations for their own information, previous to their designing or advising the construction of important works, particularly reservoirs, catchwater drains, and embankments to retain the water falling among the hills, and to store it, either for the supply of towns, for sanitary purposes, or as a means of obtaining mechanical power.

The water supply for the populous and rapidly increasing capital of the cotton manufacture, Manchester, and its neighbourhood, has led to many local observations among the ridges of hills, within reach as it were of the town (or city as it may now be termed), where it was supposed a sufficient area of high and sloping ground might be found to catch the rain, and afford the requisite water shed, from which a supply might be gathered in the rainy months, and stored for use in dry weather.

In the previous chapter we have shown that circumstances of apparently small importance may have considerable influence upon the amount of rain registered. A few indications of the errors commonly committed by observers fifteen or twenty years ago may be useful.

It may be mentioned that some of the rain-gauges used were of the old construction; wherein the rain is caught in a funnel, received in a tube, and measured by a staff or graduated rod, by dipping it into the tube and noting how much of it is wet, or by the rod rising with a float as the rain-gauge fills.

It is obvious, that when this staff was left in the instrument, projecting, as sometimes happened, considerably above the dish or funnel head of the instrument, it caught the drifting rain; and as the wind was violent or otherwise, so were the results affected: thus they varied from the neighbouring rain-gauges, which had no staff, and where the quantity of water contained in it was either read off in a graduated glass tube attached to the instrument, or poured into a separate measuring glass.

It may be said that such errors might readily have been avoided by a little more care and judgment, and something like uniformity of system in the construction and position of instruments, and the time and method of observing; but it must be remembered that such observations were generally undertaken either by individuals, or by a few persons who paid their own expenses; that the engineer was often obliged to depute his pupil or assistant to organise, among the natives of the hills, an extemporary body of observers, by whom his instructions were hardly understood, even when he flattered himself he had made them very clear and explicit.

The author has before him a series of tables, the "Collected observations, magnetic and meteorological, made throughout the extent of the empire of Russia, and published annually by order of his Imperial Majesty Nicholas the First." These observations, printed in French, are made by the engineers of the mining corps, at their several stations; and they are so copious and exact, that except the observations made at Greenwich under the Astronomer Royal, and perhaps those at the universities, few registers kept in England will bear comparison with the Russian tables. The rain-gauge used is composed of two cylindrical vases of copper, placed one above the other, and communicating by a small tube; the upper vessel is open at the top, and is larger in diameter than the under one. The rain water received in the upper vessel passes through the pipe into the closed receiver below, whence it is drawn off by a tap into a wide-mouthed tubular glass measure, graduated in equal divisions. The rain-gauge is filled with water, until it rises in the upper cylinder to a point marked beforehand, and then the water is drawn off by the tap until it has fallen an inch. It is received in the cylindrical measuring glass as it runs out of the lower copper vessel; and if it fills this glass $13\frac{1}{2}$ times, its total capacity being $\frac{1}{13\cdot5}$ of an inch, or 0·074 in depth of the upper vessel, it is only necessary to divide the glass into $7\frac{1}{2}$ parts, to show the hundredth part of an inch of rain.

Twice a day the quantity of rain or snow that falls is noted and registered, that is, at 8 o'clock, morning and evening; except in case of a heavy shower, when the time, the duration, and the quantity of rain fallen are also noticed, along with the half-day's downfall, and a corresponding remark made.

In winter the rain-gauge is taken into a warm room every twelve hours, to melt the snow or hail received in it, and a second instrument immediately put in its place; every station being provided with two rain-gauges for this purpose.

It is obvious that such instruments as these may be readily managed by private soldiers, and that correct results may be obtained without much nicety of construction, and without much regard being paid to any of the dimensions, except the divisions of the measuring glass.

A very simple apparatus has been found useful by the author, where a better could not readily be obtained. Let a tin funnel be made in the form of a shallow cylinder, or hoop, of $13\frac{17}{32}$ inches in diameter, equal in area to one superficial foot, and having the bottom conical, with a pipe of small diameter in the centre, say three quarters of an inch or thereabouts: fit this pipe into the neck of a bottle (a stone-ware bottle is better than a glass one), and let it reach nearly to the bottom of the bottle; put in a little water until it rises above the end of the pipe, to close and seal it so as to prevent loss by evaporation, and the rain-gauge is ready for use.

Weigh the whole apparatus and note the weight; then set it to receive the rain that may fall with the bottle sunk in the ground, and as the top or mouth of the funnel is one superficial foot in area, and as a cubic foot of water weighs 1,000 ounces, every ounce of water received in the bottle is one-thousandth part of a foot in depth of rain fallen.

One of the most approved rain-gauges has a funnel similar to the last, also a foot in area, mounted on a stand-pipe which receives the water, which should be carried down to the bottom by a slender tube trapped with water at the end, and the stand-pipe has a suitable base or foot. The rain received in the pipe is read off by means of a small graduated glass tube fixed to its side and communicating with it. If the stand-pipe's section be one-tenth of a superficial foot in area, and the glass be divided decimally, every inch of water in the glass will indicate one-tenth of an inch of rain; or the glass may be graduated by trial to suit the stand-pipe.

A convenient portable rain-gauge is also made, which answers sufficiently well for temporary purposes, and shows the quantity of rain fallen with tolerable accuracy. It is made of japanned tin, something like an ordinary coffee-pot, and is surmounted with a funnel, equal in area to one-fourth of a superficial foot. The water it receives is poured into a graduated measuring glass, whose diameter or section is so proportioned to the area of the funnel as to indicate the depth of rain in inches and decimals; or the glass may be marked in the same way as those used at the Russian stations before described.

It is better to use any of these modes than a measuring rod or staff, which often gives incorrect results; and the rain-gauge should be set with the mouth of the funnel exactly in a horizontal plane or level, and not exposed to blasts or eddies of the wind, nor sheltered from the ordinary breezes, so that its position may correspond with the general surface of the locality, and catch a fair average of the rain.

The rain-gauge recommended for local observations by Mr. James Glaisher, F.R.S., Secretary to the British Meteorological Society, is a simple cylindrical vessel, eight inches in diameter and thirteen inches high, sunk in the ground eight inches, and consequently having the top edge five inches above the ground. Into the top or mouth of this vessel is fitted a funnel, formed of a plain inverted cone six inches deep, the apex of the cone ending in a bent tube of an eighth of an inch in diameter, bent so as to form a trap and prevent evaporation, but yet discharging the water freely into the vessel below. For measuring, he recommends a tall cylindrical glass, holding one inch of rain as received in the gauge, and divided into ten parts, which may again be sub-divided into ten by using a scale.

It is due to Mr. Glaisher to state, that having noticed the great want of an accurately kept series of observations of atmospheric and meteoric changes in this country, he has sedulously devoted himself to the establishment of local observatories throughout the kingdom. He furnishes the observers with formularies, he directs them in the selection of instruments, and he visits their localities if needful. He thus interests others in the same useful labours he himself pursues; and he has succeeded in inducing the librarians and

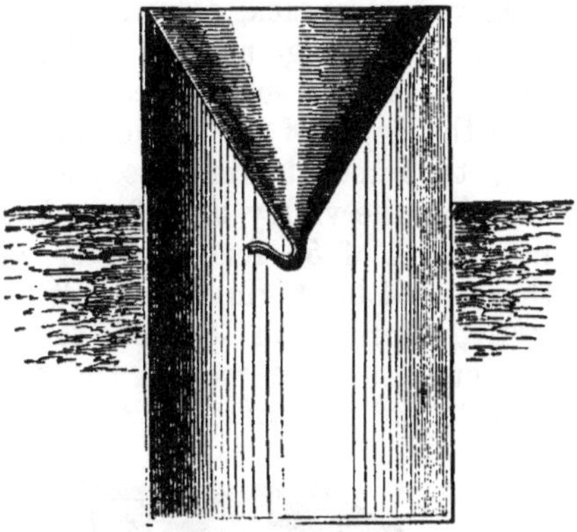

Fig. 6.—Rain Gauge.

secretaries of literary and scientific institutions, mathematical instrument makers, and private gentlemen, to institute observations of this kind on one general and uniform system. They transmit to him printed forms filled up, and he arranges and tabulates them. Such works as these cannot be too highly commended; and as the farmers' crops and land owners' rents depend on these changes, they will do well to observe them also.*

Having thus shown how the rain may be measured by the gauge and collected in the reservoir, confined by embankments, and regulated to supply the required power to turn the mill, and substitute mechanical force for manual labour, let us not forget the old and homely proverb, that "Fire and water are good servants, but they are bad masters."

The weight of a cubic foot of water being 62½ lbs., it presses with that force upon the square foot of surface on which it rests, and every foot in the depth of water adds

* Mr. Glaisher's duties as registrar of the British rainfall have been performed by Mr. Symons for some years past. Mr. Symons issues a report annually. The gauge used by the hundreds of observers who supply him with registers is somewhat different from that shown in Fig. 6: it is called the British Association gauge. It can be purchased for half a guinea.

that force to the pressure; so that if the depth of a reservoir be 54 feet, every square foot of ground at the bottom sustains a pressure of $62\frac{1}{2} \times 54 = 3{,}375$ lbs., or somewhat more than a ton and a half. What care then must be necessary in preparing a large area to resist such a weight, and render it secure under such a pressure!

But it is not only in a vertical direction that the pressure of a fluid acts: it thrusts against the embankment with a general force over the whole surface, equal on the perpendicular line to half the depth of the liquid column in a lateral direction, tending to force it from its place; and if the mound or dam crossing the valley be 600 yards in length, or 1,800 feet, with a depth of 54 feet of water against it, the whole lateral thrust will be, taking it at half the above perpendicular pressure, more than 83,000 tons; while the lateral pressure exerted at the foot of the embankment is equal, or nearly so, to the perpendicular force.

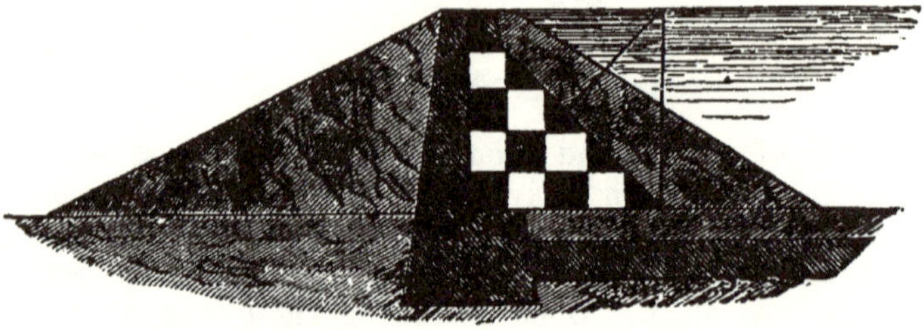

Fig. 7 Diagram to show the Pressure of Water against an Embankment.

These dimensions are not hypothetical; the author was for many years in the habit of walking, almost daily, along such an embankment, and he has seen it at times severely tested.

Let it be remembered that in such situations a bed of alluvial soil often rests upon clay, with water oozing between them, and that the beds in such situations slope in the direction of the valley, and consequently in that of the thrust. This is no uncommon case, and unless the greatest judgment and foresight be exercised in forming such dams in the outset, they may sooner or later yield to the enormous force pressing against them, and carry ruin and devastation with them.

The author has seen railway embankments, hastily raised to cross the valley, where an inclined substratum of clay, lurking as it were beneath, intercepted the surface water, and caused an insidious and slippery parting in the measures,

slide as if they had been launched for several yards, and wrinkling up the green sward, to the dismay and loss of the too clever contractors who risked such an experiment. But in making dams to confine water, nothing will justify such risk; and as this little book may fall into the hands of young practitioners in remote places, the author strongly urges that in constructing such works nothing should be left to chance; that well-constructed banks with flat slopes, stout puddle walls and lining, earthwork and masonry sufficiently massive to resist unflinching the greatest possible amount of thrust they may have at any time to sustain, should be constructed in all cases; that means of relief and discharge be provided to meet extraordinary seasons; and that both surface and substrata be carefully bored and examined before the works are commenced, and diligently watched during their progress, for great responsibility lies upon the man who attempts, with insufficient means, to restrain a destructive and overwhelming torrent.*

CHAPTER VI.

HORIZONTAL WATER-WHEELS AND SIMILAR MACHINES.

The primitive application of water-power to turn millstones has been noticed in the early part of this book, and the employment of horizontal water-wheels, with vertical axles, is still considered by French engineers to be in many cases advantageous, as presenting great simplicity and economy, both in construction, maintenance, and application; as requiring but little space, and in being able to work in floods and in frosty weather.

In driving corn-mills they need no toothed-wheel work,

* The remarks as to the construction of embankments given in the preceding chapter are very meagre and insufficient. The reader should consult Mr. Bateman's "Account of the Bann Reservoirs;" "Minutes of Proceedings of the Institute of Civil Engineers;" Mallet's account of same; "Reports on the Dodder Reservoirs;" in *Weale's Quarterly Papers;* "The Parliamentary and other Reports on the Failure of Reservoirs, such as that at Bradfield;" and the systematic works of Mahan, Sganzin, &c. There are excellent articles on Embankments in the *Edinburgh* and *Britannica Encyclopædias.*

and in besieged towns they can be worked at all times without interfering with the defences, being either placed altogether out of harm's way, or costing but little to shelter them from the enemy's fire.

Such is the opinion of experienced officers of the French artillery, and we are indebted to two of them—MM. Piobert and Tardy—for an elaborate series of experiments, and an excellent report on the useful effect of the ordinary horizontal water-wheel at present used in France. Those on which the experiments were made are at Toulouse, where the two dams (*barrages*) of the Garonne, and the abundance of water in the canal of the south, near its discharge into that river, have rendered disposable falls of water sufficient to put in motion a great number of corn-mills by means of horizontal water-wheels. These wheels are of two kinds: those situate on the rivers are called bucket-wheels (*à cuve*), and are similar to what are used at Cahors, at Metz, and other places; those which are placed on the canals are called whirl-wheels (*roues volants*), and much resemble those which have existed from time immemorial, and are turned by the percussion of the water upon curved floats, which are here used instead of the ladles that are fixed round the axles of the mills of the Alps.

It may here be remarked that in Northern Africa several rude mills are to be found in the same fashion as they have existed for ages, among a people the least advanced in the arts of industry; many of them are on the great falls of the Rummel, at Constantineh, and instead of ladles these have pieces of wood rudely driven into the upright axle, like spokes into the nave of a cart-wheel. A channel being made from the river, at an inclination of 30 or 40 degrees, the water is directed against the side of the wheel, and having done its work, it is returned to the river and employed again and again as it descends the hill to turn a series of such mills. In some of these the upper end of the vertical axle is fitted with a bent arm or crank, and the millstone, which, in such cases, is fixed in an inclined position of 10 or 15 degrees to the horizon, is forced round by it. With these mills they prepare the coarse meal, which, being cooked in steam, makes the "couscousou," the common food of the natives.

The localities at Toulouse afforded many favourable circumstances for making experiments, besides the general employment of these two kinds of wheels, so that the results of both could be readily and exactly compared by the same dynamometers and other instruments used by the same observers.

The results of all the trials appear to be that on the horizontal water-wheels with buckets, the effects produced at ordinary speeds varied from 15 to 27 per cent. of the power employed when the mills and wheels were in good condition. The speeds were varied from 60 to 135 revolutions per minute; but the best effect seems to have been obtained at about 90 revolutions, with a total fall of water, measuring the difference of level above and below the wheel of from seven to eight feet. The wheels were about five and a half feet in diameter; that of the millstones is not stated in the report, but they appear to have been such as are in general use—probably about four and a half feet. The water which drove these wheels was discharged through an ordinary sluice, and passing through a channel of stone-work, was thrown obliquely on the wheel.

The other kind of horizontal wheels experimented upon was distinguished by the name of "*roues volants*," here termed a *whirl-wheel*—for the term fly-wheel, as we now use it, is applied to a very different piece of machinery, namely, the massy cast-iron regulator of steam-engines and other heavy works.

These wheels received the water directed upon them through an inclined pyramidal trunk of wood upon one side of the wheel; the larger end of the trunk being closed, or the entrance of the water regulated by a sluice, against which was a head of water of fourteen or fifteen feet, to which the inclination of the trunk, or about two feet more, may be added, so that the ladles of the wheel were acted upon by the weight and impulse of the water, and were so formed as to continue such action until the water escaped between them, and passed through the wheel.

When these wheels made 102 and 108 revolutions per minute, the useful effect was from 29 to 33 per cent., and when the resistance of the work done reduced their speed to 90 and 85 turns per minute, their effect reached to 39 and 40 per cent. of the power expended, the useful effect of these wheels being nearly the same as that of the old undershot water-wheel.

The difference in construction between the two kinds of mills appears to be very slight, and their dimensions and cost to be the same, or nearly so; but the supply of water being abundant, the millers paid no attention to the quantity expended in performing a given amount of work. (See Fig. 9.)

The wheels are made of cast iron, and the pivot of the upright shaft stands upon a foot-bridge or lever, fixed at one end, and regulated at the other end by a second lever

Fig. 8. **Roue à Cuves.**

placed in the mill above, so that the millstones may be adjusted to grind closer or otherwise in the usual way.

It is, perhaps, impossible to construct a more simple and

less expensive corn-mill, where plenty of water is at hand to drive it, and the author has been somewhat diffuse in explaining it, as there may be many localities, especially in New

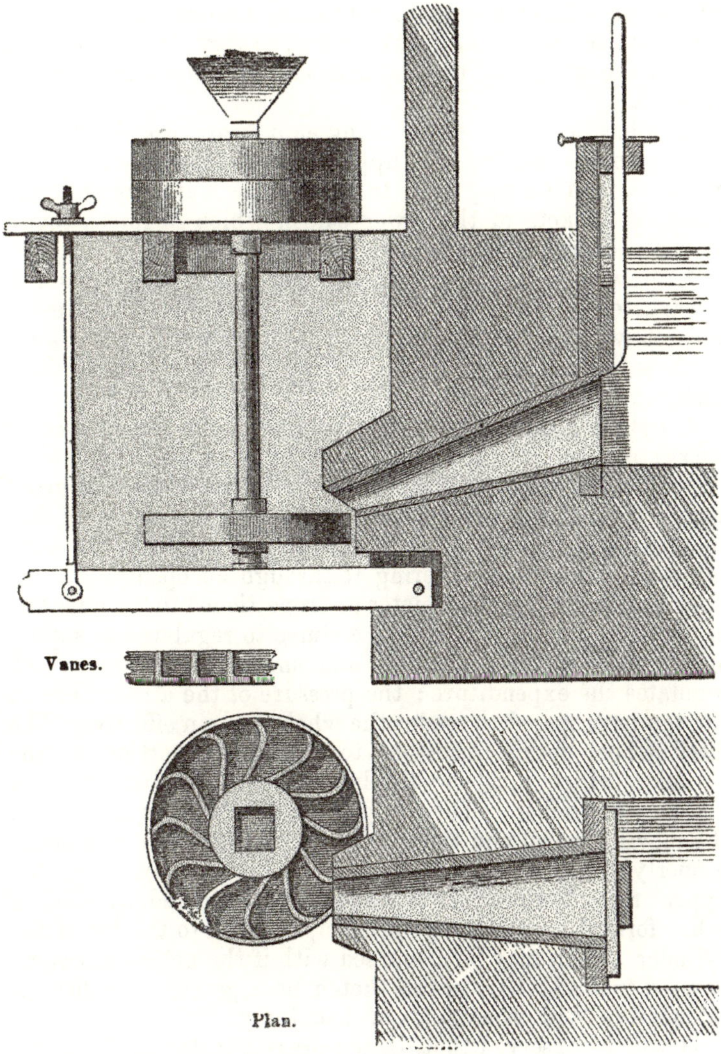

Fig. 9. Roue Volant.

Zealand and other colonies, where such simple and cheaply made mills would be of great advantage to a small community of settlers.

It is, however, well known to all experienced millwrights, that a much greater amount of useful effect is obtained when water acts by its gravity or pressure than when it acts by its impulse; and it has been found expedient in some places, where the millers desired to retain the horizontal wheel and to economise the expenditure of water, to vary its construction, so that the weight of the water should act, and that without impulse.

This has been effected by using as it were two wheels, one laid upon the other; the upper wheel being fixed and immovable, and serving only to direct the water against the vanes or buckets of the lower wheel, which is forced round by the pressure so directed against it.

A drawing of such a mill, the wheel 5 feet in diameter, constructed in Italy, has been kindly forwarded from Genoa, and shows the next improvement made in this useful machine. It is known in France and Germany as Koechlin's turbine.

A cylinder is formed of cast iron, wrought iron plates, or wood strongly hooped, and is made open at the top, unless the millstone rest upon it when the power is used to grind corn. The upper end of the cylinder is somewhat higher than the head of the column of water intended to act upon the wheel, the water entering it through an opening on one side, and the internal diameter as proportioned to the quantity of water to be used; there is a sluice to regulate the supply at top, fixed in the pentrough, and another at bottom which regulates the expenditure; the pressure of the atmosphere on the top is supposed to render the whole column effective. The fixed wheel forms a bottom to the upper portion of the cylinder, which must be firmly secured to a foundation of masonry or timber. The upright shaft or axle is fitted into the moving wheel and turns with it, passing through a collar properly bored and lined with brass, in the centre of the upper or fixed wheel; it is steadied and secured by another collar formed on a frame or bracket, screwed to the top of the cylinder, which may be dispensed with if the nether millstone be used instead; but, in the sketch here given, the wheel is intended to turn other kinds of machinery.

It will be observed that, in the present instance, the upper portion of the cylinder, above the fixed wheel, is made of cast iron, and that the lower part is made of wrought iron plates. The pressure of the water is directed by the vanes or guide-curves of the upper wheel into the buckets of the lower

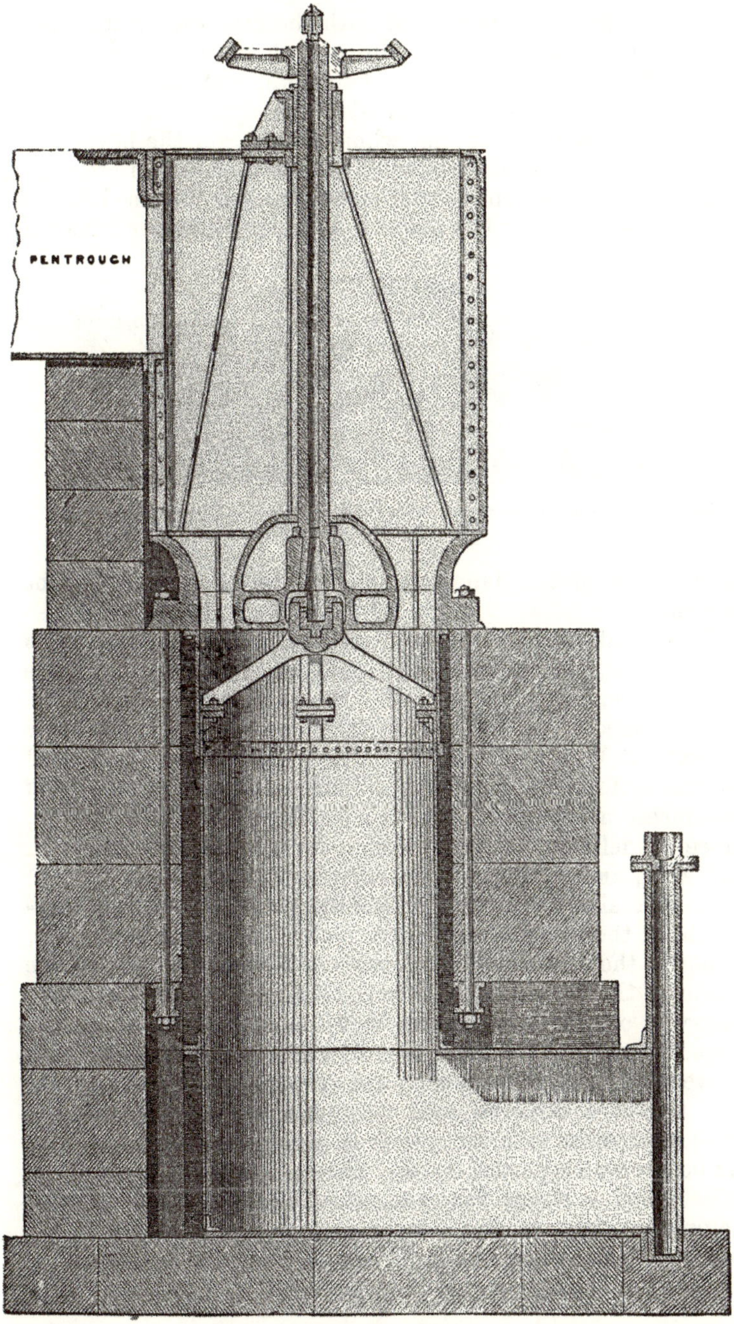

Fig 10 Sectional Elevation

one, so as to bear upon them with the greatest effect, while by the regulation of the two sluices (which are not shown in the wood-cut), the cylinder is kept full, and the descending column of water passes like an eddy through the wheels with a force proportioned to the whole height, for the lower end of the cylinder is immersed in the water, which in ordinary times just covers the outlet opening, and in flood times rises above it, so that the power due to the differenee between

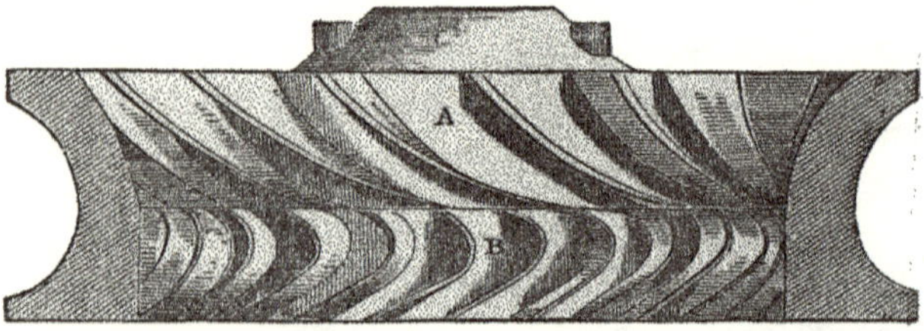

Fig. 11. Section A. The fixed part, with Guide Curves. B. The revolving part or Wheel, with its Buckets.

the surfaces of the dam and the tail water may always be available.

This combination is a great improvement upon the wheels at Toulouse, before described; and the same principles have been further developed by MM. Fromont, who obtained the Council Medal, at the Great Exhibition of 1851, for their Horizontal Water-Mill of fifty-five horse-power, constructed on the system of M. Fontaine Baron—a machine capable of the nicest adjustment, and applicable to any manufacture, however delicate, as the spinning of silk and fine yarns of any kind, and giving a useful effect of 75 per cent.

In several respects this resembles the last-described, inasmuch as there are two wheels placed in the same manner, either at the lower end of a large iron cylinder, open at the top or at the bottom of a cistern of strong woodwork. The reservoir or cistern intended to be used for this wheel was about 8 feet high, but it was not exhibited, as it would have prevented a complete view of the interior arrangements. It was therefore removed, and the upper work regulating the sluices, together with the governor, were placed on a cast-iron frame above, supported by six pillars. The upper, or fixed wheel, had a strong flange surrounding it, and cast with it, having provision made for securing it with screw-bolts to the bottom of the cistern or tank, and beneath the revolving, or under wheel (in this case not included in a cylinder, as in

Koechlin's plan); a clear way of escape for the effluent water,

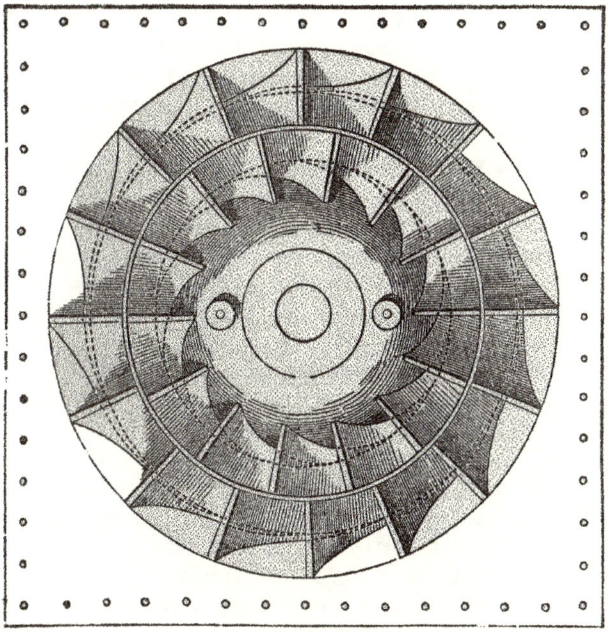

Fig. 12. Plan of fixed part, with Guide Curves.

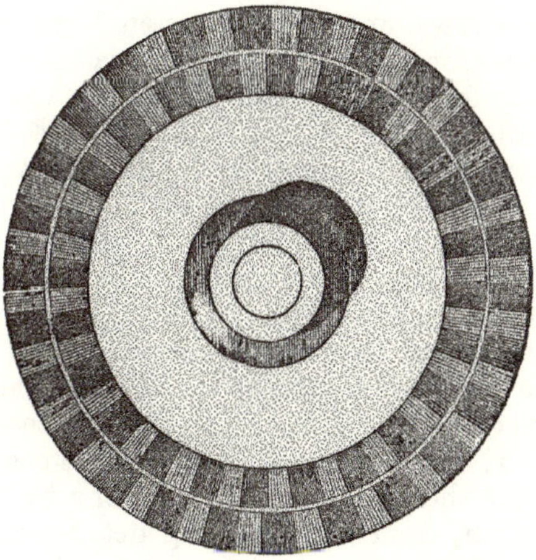

Fig. 13. Plan of revolving part, with Buckets.

after it had done its duty, was left on all sides. The fixed wheel, thus attached by its flange to the bottom of the tank, was about 6½ feet in diameter, and had a strong cast-iron pipe, about a foot in diameter, fixed by bolts to its centre, and rising up above the level of the cylinder or cistern, where it was secured by a strong frame, serving to bind together the whole fabric, and to support the axis of the revolving wheel which passed through the pipe and worked in a gun-metal collar above.

In the fixed wheel there were forty-two openings disposed in two concentric circles; the outer circle having twenty-four, and the inner circle eighteen; these openings radiated from the centre of the wheel, and were formed so as to direct the water into the buckets of the revolving wheel below it. The outside diameter of the outer circle of openings was about 6½ feet, and the inside diameter of the inner circle about 3½ feet; there was a division, or partition, of an inch in thickness, between the two circular series of openings, so that they might be about 8½ inches long.

Each of these openings was fitted with a sluice made of cast iron, with the back of it properly curved in a vertical direction, so as to direct the water in passing through the openings and lead it fairly into the buckets.

Every one of these sluices is suspended and attached to a wrought-iron rod, about an inch in diameter, screwed into the sluice, and standing up about 18 inches. On the top of these rods is secured a strong flat ring; the rods are fitted with collars, screws, and nuts to fasten them into the ring, or rather rings, for there are two; that is to say, one for the outer, and one for the inner circle of sluices and openings.

Each of these rings is suspended by three strong round rods, with regulating screws cut upon them, so that by means of wheel-work the two circles of sluices can be simultaneously raised and opened, or lowered and shut, at pleasure; or the openings may be enlarged or contracted, to regulate the speed of the machine, by adjusting the discharge of water through them; and this regulation is effected by a conical pendulum or governor similar to that of the steam-engine.

This apparatus of MM. Fromont has been improperly termed a turbine, by which it is understood some wheel resembling the shell so called, and working by re-action; but it will be clear, from the foregoing details, that it was, in fact, a horizontal water-wheel, in the construction of which great ingenuity and mechanical skill were displayed.

In awarding the Council Medal to this machine, the jury mentioned the advantages possessed by it, and by similar

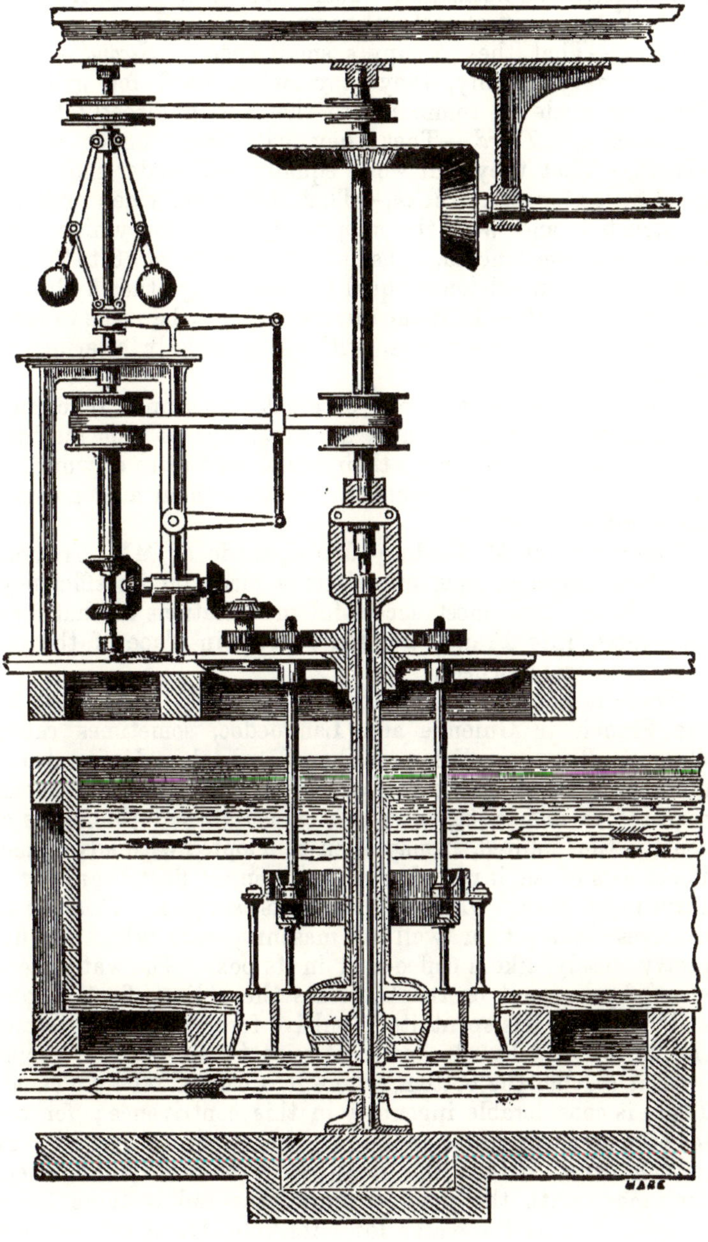

Fig. 14. Sectional Elevation.

well-constructed horizontal wheels and turbines, of frequent use in France, but almost unknown in England.

First,—That they occupy a small space. *Second*,—That turning very rapidly, they may, when used for grinding flour, be made to communicate the motion directly to the mill-stones. *Third*,—That they will work under water. *Fourth*,—That they will work equally well under small and great falls of water. *Fifth*,—That they yield when properly constructed, and with the supply of water for which they have been constructed, a useful effect of from 68 to 70 per cent.; being an efficiency equal to that of any other hydraulic machine. *Sixth*,—That the same wheel may be made to work at very different velocities, without materially altering its usual effect.

This last property is one of great importance in certain applications, and constitutes an advantage of this machine over most others that have their established rates of working, from which it is not possible to deviate without a proportionate sacrifice of power.

The wheel of M. Fontaine Baron, made by MM. Fromont and Son, appears, from experiments made on its efficiency, to be one of the most successful modifications of that form introduced into France by Fourneyron, in place of the old horizontal water-wheel.

There is another simple and useful water-wheel used by the French in Guienne and Languedoc, sometimes called *Roue à Poire*, or the pear-shaped wheel. It is also a horizontal wheel with a vertical axis; and when the power required is not great, the water plentiful, and the means of construction limited, it may often be adopted with advantage. It consists of an inverted cone, with spiral float-boards of a curvilinear form, winding round its surface. This wheel revolves on a pit or well of masonry, into which it fits pretty closely, like a coffee-mill in its box. The water, conveyed by a spout or trunk, strikes the oblique float-boards, and, when it has spent its impulsive force, it descends along the spiral float-boards and continues to aid by its weight until it reaches the bottom, where it is carried off by a canal. There is considerable ingenuity in this contrivance; for the jet of water being first applied to the upper, or largest part of the cone, strikes the float-boards at the point where they move with the greatest speed, the radius there being longest; but as the water loses its velocity, in consequence of the motion it has imparted to the wheel, it descends in the cone and acts upon the floats lower down, where, the radius being less, they move more slowly; and the water is still beneficially employed until it quits the wheel.

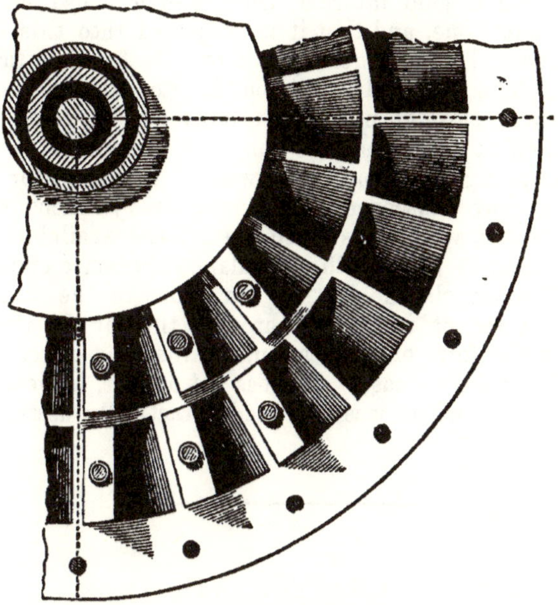

Fig. 15. Plan (one-fourth) of the fixed part, showing the Guide Curves and six of the Sluices or Regulators.

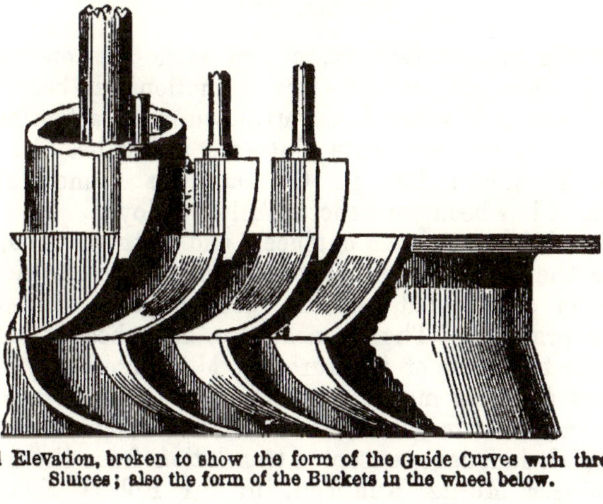

Sectional Elevation, broken to show the form of the Guide Curves with three of the Sluices; also the form of the Buckets in the wheel below.

It is, perhaps, not necessary to describe the Danaide of M. Dectot, nor the Conchoidal-wheel of Euler, as neither of these are in use, excepting so far as to notice that, in the first of these, the water acts in a narrow annular channel, at the circumference of the machine, formed between a hollow external and a solid internal drum, which is also shallower than the outer one, and that it is projected into this hollow ring or channel by inclined spouts or jets forming tangents to its circle, so as to drive it round, after which it goes out at the centre, through passages suitably formed, so as to continue the effect of the water as long as it remains in the wheel.

In designing the conchoidal wheel, the inventor seems almost to have discovered the reactionary effect of the effluent water, which has since been made available in the turbine, that is to say, the unbalanced pressure opposite to the orifices impels the wheel, somewhat in the same way that a sky-rocket is driven through the air in its rapid flight by the force or recoil acting against the closed end of the rocket tube or case. It does not appear, however, that this horizontal wheel was ever made practically useful.

CHAPTER VII.

TURBINES AND THE REACTION OF WATER PRESSURE ON WHEELS, ETC.

Within the present century, a new mode of applying the power of water to produce circular motion has been introduced, and of late years it has attracted much attention, and received many improvements; hitherto it has been but little used in England; but in Germany, in France, and in America, it has been very successfully employed.

It is chiefly to German engineers and mathematicians that we owe the investigation of the principles on which the turbine is constructed, and the best methods of reducing them to practice. The French, also, have been prompt to appreciate the value of the turbine; M. Arago has given his testimony as to its merits, and other French writers of note have examined it in detail; but we have no complete work in the English language, except a translation from the German of Professor Moritz Ruhlman, by Sir Robert Kane, and rendered valuable to practical men by his observations.

In order that the subject of this chapter may be better

elucidated, and traced from its first elements, it may be proper to notice the philosophical toy which figures in many works on hydrostatics and hydraulics, as Dr. Barker's mill, but is by most persons passed over as a mere plaything, useless for practical purposes; it involves, however, principles of action which, when well and scientifically carried out, lead to most important results.

Dr. Barker's mill consists of an upright pipe or tube, with a funnel-shaped open top, but closed at the lower end; and from the lower end project two horizontal pipes or arms, also closed at the outer ends, and placed opposite to each other, at right angles with the vertical tube, so as to form a cross. Near to the end of each horizontal pipe, and on one side of it, is a round hole, the two holes being opposite to each other. The upright pipe is mounted upon an axis or spindle, and is kept full of water flowing into the top.

The water, issuing from the holes on the opposite sides of the horizontal arms, causes the machine to revolve rapidly on its axis, with a velocity nearly equal to that of the effluent water, and with a force proportionate to the hydrostatic pressure given by the vertical column, and to the areas of the apertures; for there is no solid surface at the hole on which the lateral pressure can be exerted, while it acts with its full force on the opposite side of the area. (See Fig. 16.)

This unbalanced pressure, according to Dr. Robison, is equal to the weight of a column having the orifice for its base, and twice the depth under the surface of the water in the trunks for its height.

This measure of the height may seem odd, because if the orifice were shut, the pressure on it is the weight of a column reaching from the surface. But when it is open, the water issues with nearly the velocity acquired by falling from the surface, and the quantity of motion produced is that of a column of twice this length, moving with this velocity. This is actually produced by the pressure of the fluid, and must, therefore, be accompanied by an equal reaction.

When the machine, constructed exactly as before described, moves round, the water which issues descends on the vertical trunk, and then, moving along the horizontal bars, partakes of the circular motion.

This excites a centrifugal force, which is exerted against the ends of the arms by the intervention of the fluid.

The whole fluid is subjected to this pressure, increasing for every section across the arm, in the proportion of its distance from the axis; and every particle is pressed with the accumulated centrifugal forces of all the sections that are nearer to the axis; every section, therefore, sustains an equal pressure, proportional to the square of its distance from the axis. This increases the velocity of efflux, and consequently the velocity of revolution; their mutual co-operation would seem to terminate in an infinite velocity of both motions; but

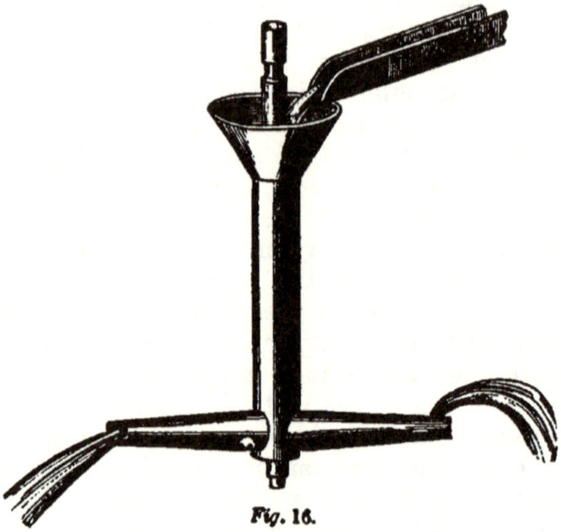

Fig. 16.

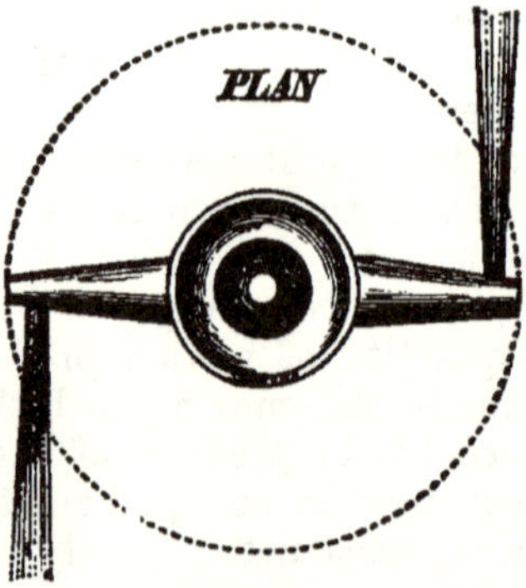

Fig. 17.

on the other hand, this circular motion must be given to every particle of water, as it enters the horizontal arm. This can only be done by the motion already in the arm, and at its expense. Thus there must be a velocity which cannot be over-passed, even by an unloaded machine. But it is also plain, that by making the horizontal arms very capacious, the motion of the water to the jet may be made very slow, and much of this diminution of circular motion prevented.

Dr. Désaguliers, Euler, John Bernoulli, and M. Mathon de la Cour, have treated of this machine; and the latter author proposes to bring down a large pipe from an elevated reservoir, to bend the lower part of it upwards, and to introduce into it a short pipe with two arms like Dr. Barker's mill reversed, and revolving on an upright spindle in the same manner; the joint between the two pipes being so contrived as to admit of a free circular motion without much loss of water. By this arrangement, a fall or column of water of any height, however great, may be rendered available. This arrangement was proposed in 1775. Some few years ago, Mr. James Whitelaw, of Paisley, attempted the improvement of this machine, and took a patent for his improvement, of which he published an account in 1845. This would seem to consist chiefly of the modifications recommended by Dr. Robison and M. Mathon de la Cour, and of the bending of the two horizontal arms into the form of the capital letter S: the water being discharged from the ends of the arms, in the direction of the circle traced by their revolution or in that of a tangent to it. The curvature is that of an Archimedean spiral, with the extremity of the arm or jet piece continued for a short distance in a circular curve, coincident with the circle described by the end of the arm. The utility of this continuation, however, seems to be questionable.

The wood-cuts show the method of striking the spiral curves to form the arm and a section of the machine.

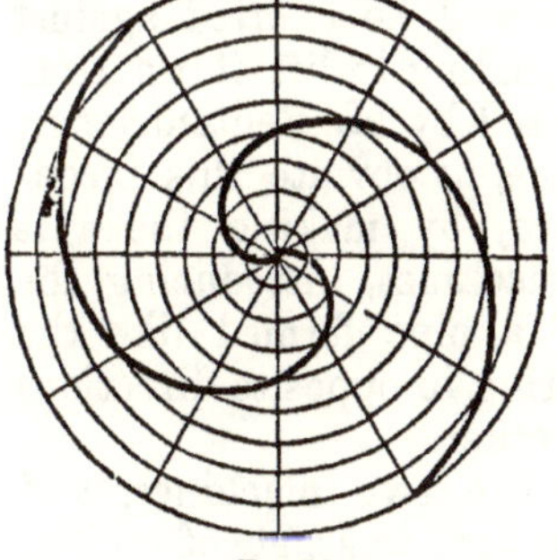

Fig. 18.

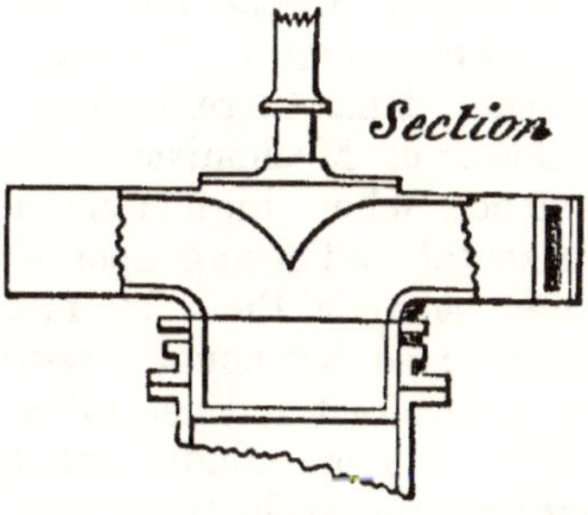

Fig. 19.

The capacity of the hollow arms is increased as they approach the centre of rotation, so as to contain a quantity of water at every section of the arm inversely proportionate to its velocity at that section, so that little of the centrifugal force may be lost. The engravings show an elevation and plan of Mr. Whitelaw's machine, and the method he proposes for forming the arms. The curvature is that of an Archimedean spiral, with the extremity of the arm or jet piece continued for a short distance on a circular curve. The utility of this continuation, however, seems somewhat doubtful.

The sections of the arms are everywhere parallelograms of equal depth, but of breadth increasing from the jet at the extremity of the arm to the centre of the machine. (See fig. 20.)

A model water-mill of the form shown in these figures, working with a fall of 10 feet, the diameter of the circle described by the arms being 15 inches, and the aperture of each jet 2·4 inches in depth by ·6 of an inch in width, the area of each orifice being 1·44 inches; the expenditure of water was 38 cubic feet, the velocity 387 revolutions per minute, and it gave an effect equal to 73·6 per cent. of the power employed.

Mr. Whitelaw states that the machine is most effective when the jets or ends of the arms revolve with a velocity equal to that acquired by a heavy body falling through the height of the vertical column.

The model used in this instance was well made and the experiments carefully conducted, but Mr. Whitelaw states that he has obtained nearly equal results in actual practice on the large scale.

It will be observed that in the employment of water of considerable altitude, a great force will be exerted against the moving part of the machine, tending to lift it up from its seat; it has therefore been proposed by M. Redtenbacher, Professor of Mechanism at Carlsruhe, to obviate this inconvenience when high falls are used, by making the axis horizontal and fixing upon it two machines, introducing the water between them by means of a pipe formed like the letter T; so that equal pressures acting in opposite directions may counteract and neutralise each other.

The axis or spindle connecting the two machines, and passing through the transverse pipe is kept in a state of tension by the diverging forces. (See Fig. 21.)

The author was indebted to Professor Wedding, of Berlin, for the first authentic information he received respecting the

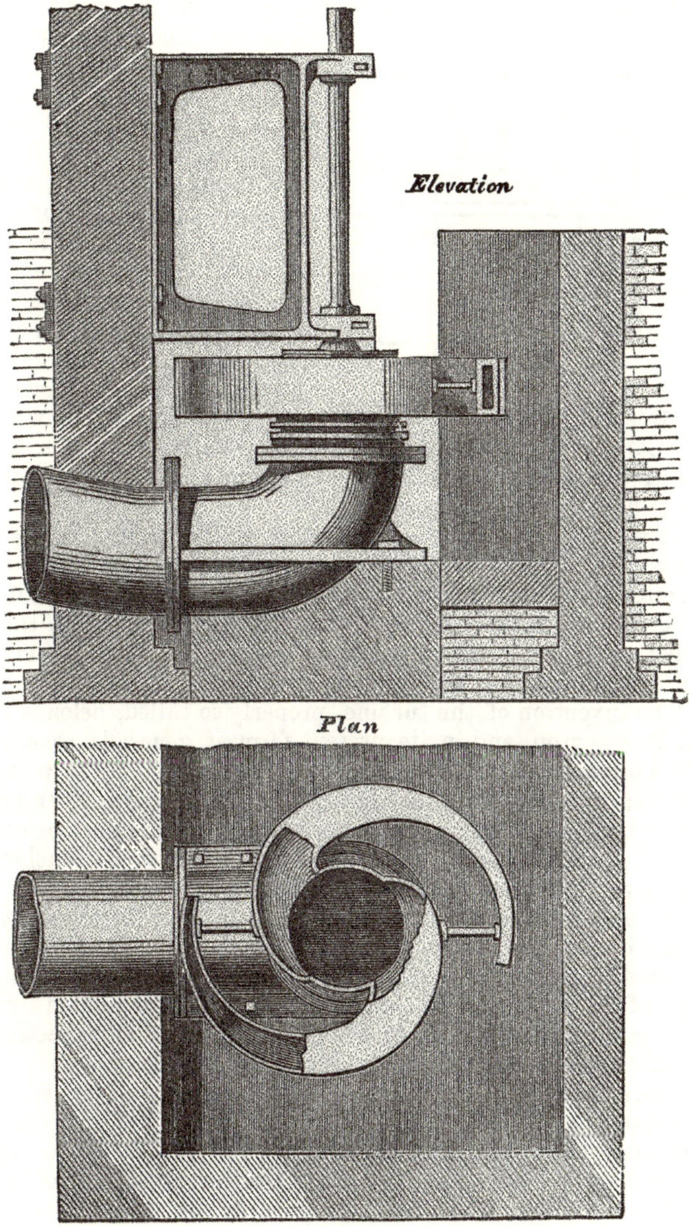

Fig. 20.

turbine, as also for a sketch and the dimensions of one of these machines, then recently erected under M. Wedding's direction. The fall, or rather the head, of water acting upon it was 20 feet, the diameter of the wheel 3 feet 8 inches, the speed 115 revolutions per minute, and the power equal to 42 horses, which drive eight pairs of ordinary millstones, with all the requisite dressing machinery. Professor Wedding afterwards sent the author from Berlin a treatise on this subject, which he had written in conjunction with M. Carliczeck, wherein several other similar machines are described, and the useful effect of the water is stated at from 68 to 80 per cent. (See Fig. 22.)

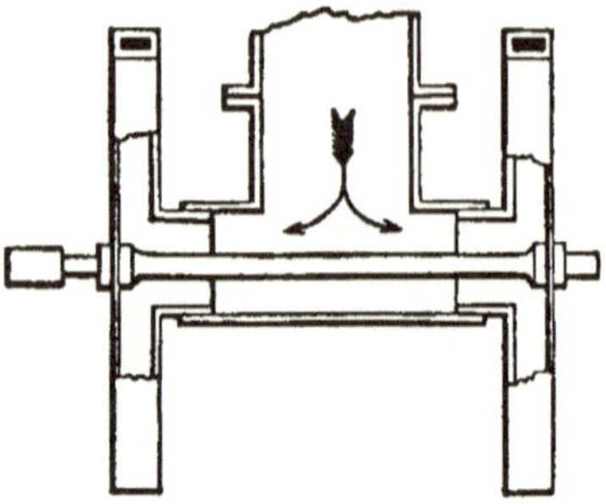

Fig. 21.

It will be seen that the great value of the turbine consists in its being applicable to falls of water so high or so low that an ordinary water-wheel cannot be used; and also that in falls of great height the velocity of the machine is so rapid, that when applied to drive spinning machinery, it needs no mill work, or but very little, to bring it to the requisite speed.

The invention of the turbine, properly so called, belongs to M. Fourneyron, and in its present form it generally consists of a horizontal water-wheel, in the centre of which the water enters; diverging from the centre in every direction, it enters every bucket at once, and escapes at the circumference or external periphery of the wheel. The water acts on the buckets of the revolving wheel with a pressure proportionate to the vertical column or height of the fall, and it is led or directed into these buckets by stationary guide curves, placed upon and secured to a fixed platform, within the circle of the revolving part of the machine. The efflux of the water is regulated by a hollow cylindrical sluice, to which a number of stops, acting simultaneously between the guide curves, are fixed.

With this short cylinder or hoop, they are all raised or lowered together by means of screws communicating with a regulator or governor, so that the opening of the sluice and stops may be increased or diminished in proportion as the velocity of the wheel may require to be accelerated or retarded.

This cylindrical sluice alone might serve to regulate the efflux of the water, but the stops serve to steady and support the guide-curves, and prevent tremor.

Turbines may be considered as divided into two classes,—the low pressure and the high pressure. The engravings will better explain the construction of these machines and the arrangement of their parts than a longer verbal description.

M. Fourneyron, who began his experiments in 1823, erected his first turbine in 1827, at Pont sur l'Ognon, in France. The result far exceeded his expectations, but he had much prejudice to contend with, and it was not until 1834 that he constructed another, in Franche Comté, at the iron-works of M. Caron, to blow a furnace. It was of seven or eight horse-power, and worked at times with a fall of only nine inches. Its performance was so satisfactory that the same proprietor had afterwards another of fifty horse-power erected, to replace two water-wheels which, together, were equal to thirty horse-power.

The fall of water was 4 feet 3 inches, and the useful effect, varied with the head and the immersion of the turbine, from 65 to 80 per cent.

Several others were now erected: two for falls of seven feet; one at Inval, near Gisors, for a fall of 6 feet 6 inches, the power being nearly forty-horse, on the river Epté, expending 35 cubic feet of water per second, the useful effect being 71 per cent. of the force employed.

One with a fall of 63 feet gave 75 per cent.; and when it had the full head or column for which it was constructed,—namely, 79 feet,—its usual effect is said to have reached 87 per cent. of the power expended.

Another, with 126 feet, gave 81 per cent.; and one with 144 feet fall, gave 80 per cent.

At the instance of M. Arago, a commission of inquiry was instituted by the Government of France, for examining the turbine of Inval, near Paris, the total fall of water being 6 feet 6 inches, as has been before mentioned. By putting a dam in the river, below the turbine, so as to raise the tail water, and diminish the head to 3 feet 9 inches, the effect was still equal to 70 per cent.; with the head diminished to 2 feet, the effect was 64 per cent.; and when the head was reduced to 10 inches, it gave 58 per cent. of the power expended, notwithstanding the great immersion of the machine.

In the year 1837, M. Fourneyron erected a turbine at St. Blasier (St. Blaise), in the Black Forest of Baden, for a fall or column of water of 72 feet (22 mètres). The wheel is made of cast-iron, with wrought-iron brackets; it is about 20 inches in diameter, and weighs about 105 lbs.; it is said to be equal to fifty-six horse-power, and to give a useful effect equal to 70 or 75 per cent. of the water power employed. It drives a spinning-mill belonging to M. d'Eichtal. A second turbine, at the same establishment, is worked by a column of water of 108 mètres, or 354 feet high, which is brought into the machine by cast-iron pipes of 18 inches diameter of the local measure, or about 16½ inches English. The diameter of the water-wheel is 14¼, or about 13 inches English, and it is said to expend a cubic foot of water per second; probably the expenditure may be somewhat more than this.

The width of the water-wheel across the pier is ·225, or less than a quarter of an inch. It makes from 2,200 to 2,300 revolutions per minute; and on the end of the spindle or upright shaft of the turbine is a bevilled pinion, of nineteen teeth, working into two wheels, on the right and left, each of which has 300 teeth. These give motion to the machinery of the factory, and drive 8,000 water spindles, roving frames, carding engines, cleansers, and other accessories. The useful effect is reported to be from 80 to 85 per cent. of the theoretical water power. The water is filtered at the reservoir before it enters the conduit pipes; and it is important to notice this, since the apertures of discharge in the wheel are so small as to be easily obstructed or choked.

The water enters the buckets in the direction of the tangent to the last element of the guide-curves, which is a tangent to the first element of the curved buckets. The water ought to press steadily against the curved buckets, entering them without shock or impulse, and quitting them without velocity, in order to obtain the greatest useful effect; otherwise a portion of the water's power must be wasted or expended, without producing useful effect on the wheel. The engravings show a section of this turbine, and a quadrant of its water-wheel, drawn to a scale of one half of the actual size. (Fig. 24.)

It is difficult to imagine that a machine so small as this can give motion to the works of a cotton mill on so large a scale. Professor Ruhlmann says, that when he saw it actually doing so, he could not for some time credit the

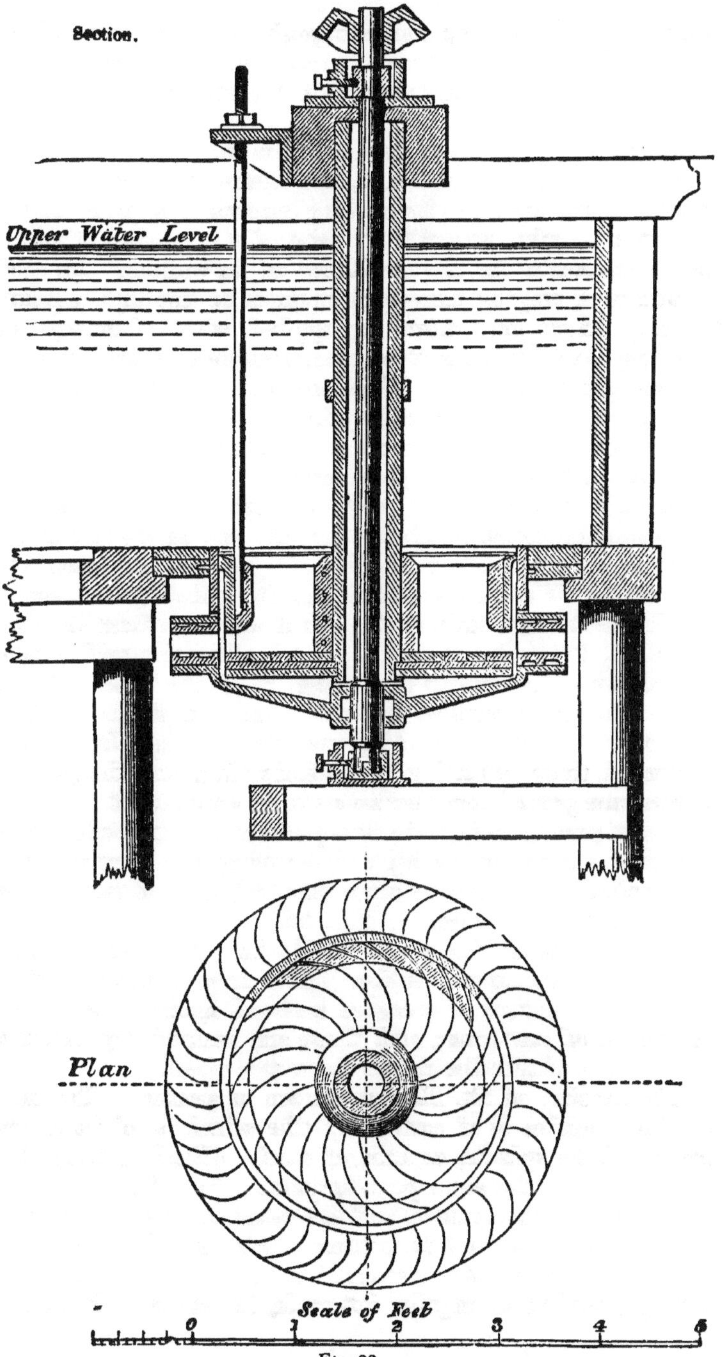

Fig. 22.

evidence of his senses; and, although he went purposely to examine it, his astonishment prevented him from comprehending, in the first instance, that the fact was really as it appeared.

There are many places, especially in hilly districts, where high falls of water are found, and where the nature of the ground affords facilities for making reservoirs, so as to ensure a constant supply, where the height of the column of water may compensate for the smallness of its volume. In such situations it may be conveyed in pipes to the high-pressure turbine, which may often be applied with advantage for grinding corn, working thrashing machines, or for crushing ore, and other purposes. There are other situations in which a great volume of water rolls, with but little fall, and it has been shown that, with a head of only nine inches, the low-pressure turbine has done good service.

The illustrations here given, as explanatory of the progress and construction of the turbine, are an elevation and plan of Dr. Barker's mill (Figs. 16, 17); an elevation and plan of Mr. Whitelaw's mill (Fig. 20); and diagrams, showing the method of striking the spiral curves to form the arms (Figs. 18, 19); with a section, showing how the mill is connected with the supply-pipe (Fig. 21); the loose collar is pressed upwards against the revolving part of the machine by three box springs, fixed between the flanges; the collar is prevented from revolving by a steady pin; and the parts in contact are ground together so as to be water-tight.

A sectional elevation of a low-pressure turbine, with one of the three screws for raising and lowering the circular sluice (with plan). The screws are connected, and act together, by means of toothed wheels. (Fig. 22.)

Also a plan of the water-wheel, the guide curves, and a portion of the circular sluice. The curved buckets, which are made of thin plate-iron, are screwed against loose blocks or pieces of cast-iron; and these are secured by means of screw-bolts within the rim of the water-wheel.

The turbine of St. Blasier, shown in section. The body of the machine is of cast-iron; the wheel is of hammered iron; and the spindle, or axis, of steel. (See Fig. 23.) The letters refer to the same parts on plan. (See Fig. 24.)

The foot of the spindle, and the pivot and step on which it revolves, are tempered to extreme hardness. The oil-pipe at the foot of the pivot is connected with a small force-pump or syringe, which, at regular intervals, injects a little oil into

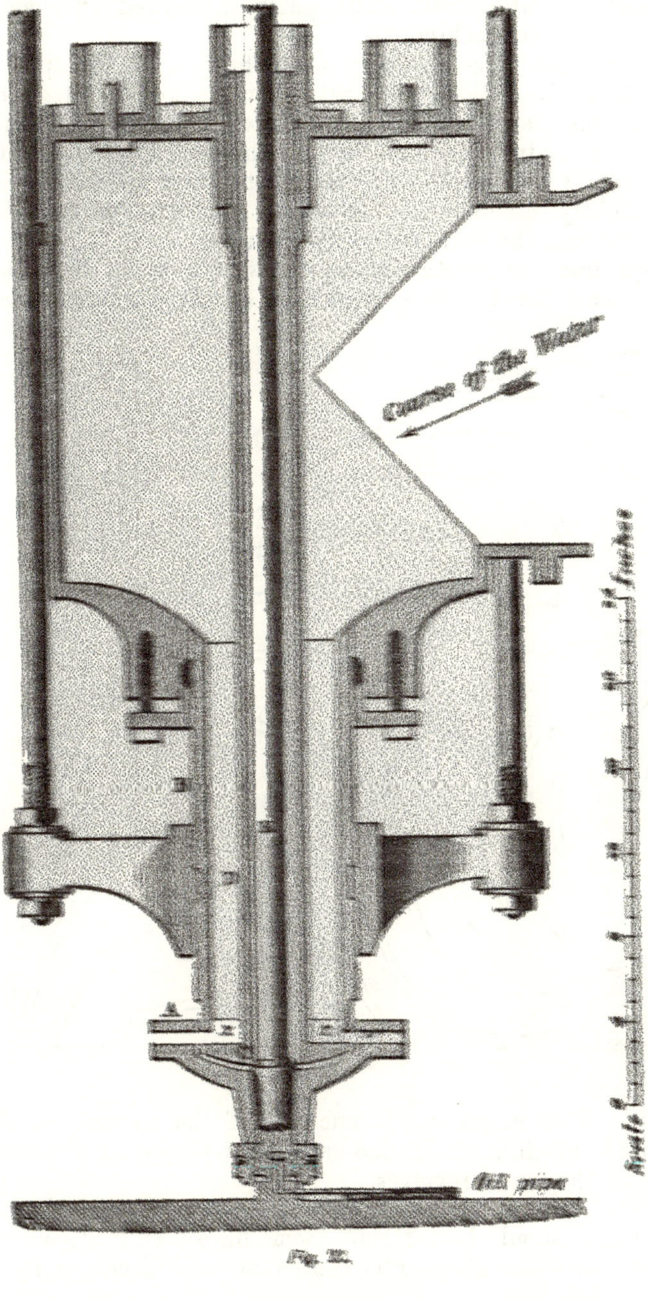

Fig. [illegible]

the step for lubrication. The pump is worked by a slow motion from the machinery.

In all cases it is necessary that the foot of the spindle shall be made hollow, and run upon a fixed pivot. The spindle must never run in a hollow step. The pivot should be quite cylindrical, and it should truly fit the spindle, with as little play as possible; the top of the pivot should be but very slightly convex. The water and mud must be carefully excluded, and the parts regularly oiled

A quadrant or fourth part of the wheel, with the guide curves, and the sluice or regulator of the turbine of St.

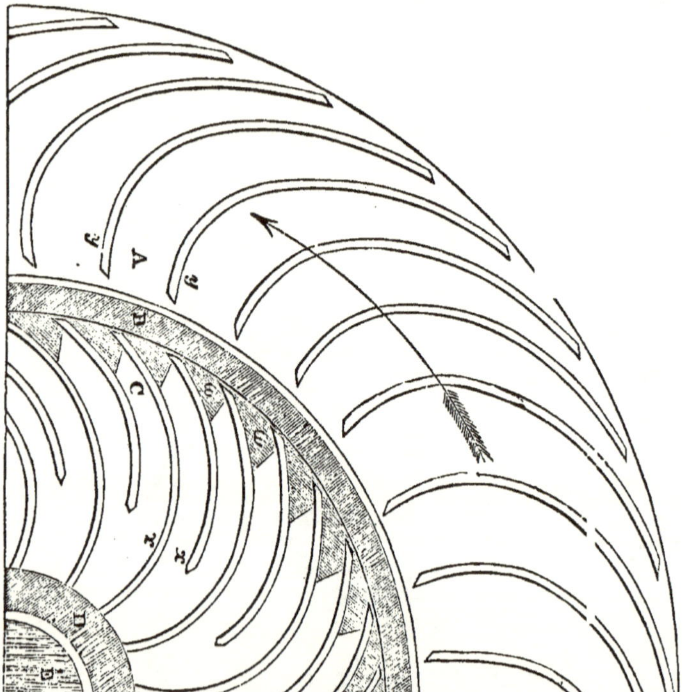

Fig. 24.

Blasier. This engraving is one-half of the natural or full size of the machine itself; the bent arrow shows the direction in which the wheel revolves.

There is also another difference in the construction of turbines which should be noticed. Some have been made which receive the water at their circumference and discharge it at

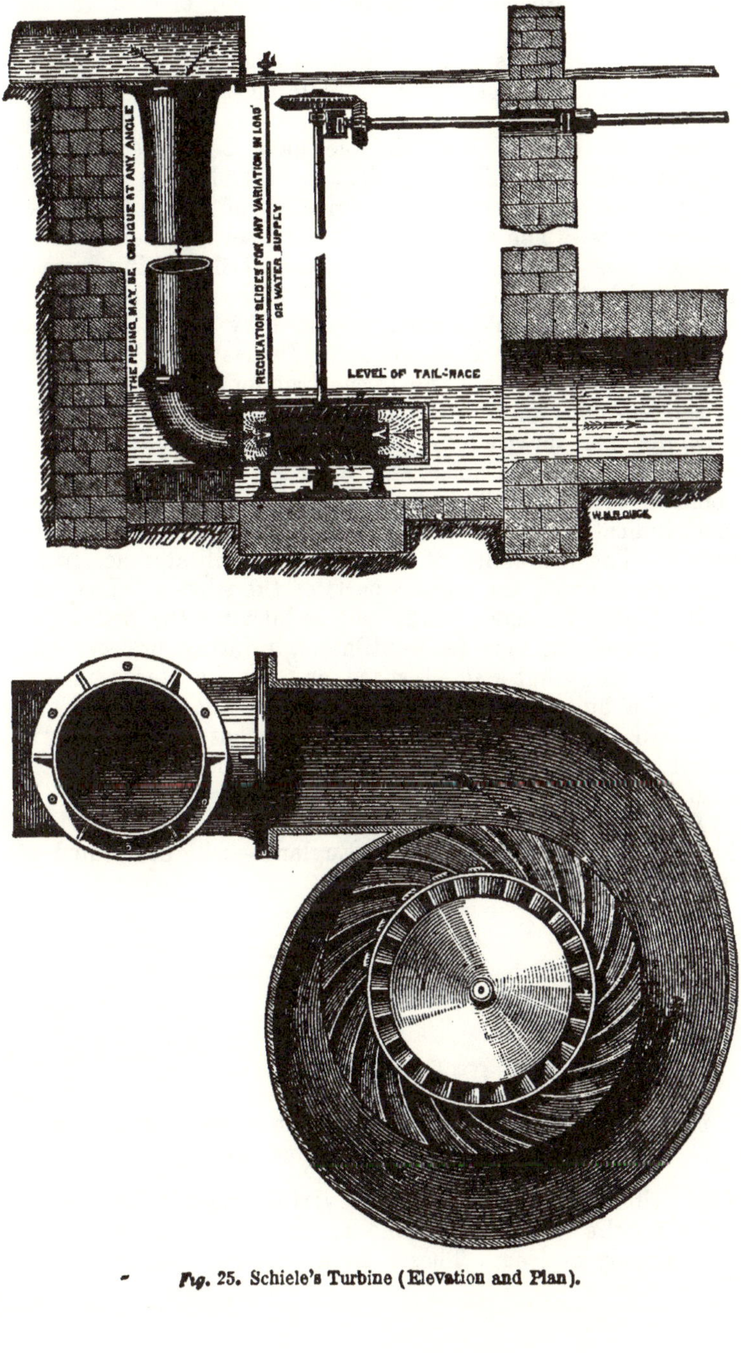

Fig. 25. Schiele's Turbine (Elevation and Plan).

the centre. Several of these have been erected in the United States; but the amount of duty done by a given quantity of water is not so great as when it is admitted at the centre and escapes at the circumference, where it can do so more freely.

Some wheels on this principle have been erected near Belfast, with good effect, by Mr. James Thompson, who read a paper respecting them at the meeting of the British Association there. The velocity of these wheels at their circumference was equal to that of a heavy body falling from a height of half the vertical column of water upon the wheel.

A compact cheap turbine, invented by Mr. Schiele, has recently come into extensive use. It has a high effective working power, and is simple in construction. The water enters at the periphery of the wheel and passes out from the sides, upwards and downwards (as shown in Fig. 25), without producing so much noise and foam as usually occur with common turbines. This freedom from noise and splashing is an indication that most of the power is taken off from the water. The simplicity of construction facilitates access to, and separation of, the various parts of the wheels. They are well adapted for small work, such as blowing the bellows of organs, working fans for ventilation, printing presses, &c., as they occupy but a small space, and can be placed in any part of a building. A turbine equal to the power of two men will require a space of about 5 inches square. A sectional plan and elevation of this machine are given in Fig. 25.

The reciprocating water-pressure engine—a machine which has been too much neglected in England—will be treated of in a future chapter.

CHAPTER VIII.

UNDERSHOT WATER-WHEELS, THEIR USEFULNESS IN A NEW COUNTRY TO SAW TIMBER, ETC.

THE primitive form and use of vertical wheels for raising water for the irrigation of land in China and the East, has been already noticed. These, simply dipping their float into a river, were turned by the current with such velocity and force as the stream might impart to them.

Yet, before quitting this part of the subject, it may be proper to mention two modes of applying these wheels which have been practised in America.

One of them was to place a strong axle across a boat, or some other vessel, of large dimensions, with a water-wheel at each end of this axle, like the paddle-wheels of a steamboat; and this vessel being moored in a current, the wheels revolved, and gave motion to mill-stones, and machinery for grinding and dressing flour, on board the floating mill. The other was by means of a similar axle and a pair of wheels, thus mounted in a boat, to cause the boat so fitted to warp itself, and to tow other boats up a rapid, by winding one end of a rope round the axle, the other end being made fast to an anchor, or other mooring, above the rapid. This means of ascending rapids in the American rivers has been generally superseded by the employment of powerful steamboats; but it is worthy of being recorded as an ingenious contrivance to derive from the existing medium itself a power to overcome it, by duly proportioning the diameters of the wheels and axles.

The next improvement was an important one; and it rendered the vertical water-wheel a powerful mechanical agent.

By penning back the stream with a dam or barrier, thrown across its channel, so as to accumulate and raise the water to a head; and by cutting a canal, or water-course, in the bank, communicating with the reservoir so formed, and re-entering the river by its side at a lower level; by erecting the wheel in this water-course, and by interposing a sluice between the wheel and the pent-up water, so as to stop or regulate its efflux, the whole power of the water, heretofore spread over the bed of the river, might be concentrated against the wheel, rushing through the opening of the sluice with a velocity and impulse due to its head and volume, and acting upon the float-boards with an amount of force and effect which could not be obtained in the open river; the water being now confined between walls of solid masonry almost in contact with the wheel, and within which it revolved.

These walls also served to support the axis of the wheel, and to retain the sluice; while a pavement of heavy stones below, between the walls, prevented the water from escaping beneath the wheel until it had done its duty.

When the sluice was shut down and the wheel stood still,

until the dam was filled to overflowing, the water passed over the barriers, or weir, and rolled on, as before, through its old channel in the river, or was discharged into it through a waste-water sluice, sometimes made self-acting by means of a balanced float, or some similar contrivance; and, on adapting such apparatus, great ingenuity has often been displayed, especially in the Shaw's Water-works, already mentioned, as well as by some of the French engineers.

Arrangements like these, so simple, so effective, and so easily made and managed, rendered the UNDERSHOT WHEEL most useful and valuable as a means of obtaining mechanical power sufficient to drive extensive flour-mills, fulling-mills, and forges, for which purposes it was, in the first instance, chiefly used, to aid an agricultural population in more readily supplying themselves with bread, woollen cloth, and iron—the principal requirements of a primitive community, with whom spinning and weaving were as yet domestic employments.

The sluice, which regulated or closed the opening through which the water issued against the wheel, was placed as low as possible, allowing only sufficient fall or space for it to escape freely after it had passed the wheel, so that it might not, as a millwright would say, "be working in back water." The opening, or the sluice, corresponded in size, or nearly so, with the width of the float-boards of the wheel, and the head of water above it caused the stream, or rather the jet, to strike the float-boards with a proportionate momentum, the wheel receiving, as it were, a succession of impulses as the float-boards continually entered the water.

Mr. John Smeaton, the most experienced and eminent engineer of his time, made a series of careful and elaborate experiments on the power and effect of water employed to turn mills, by means of undershot and overshot wheels. Accounts of these experiments, with all their details and results, were submitted by him to the Royal Society, and printed in their Transactions; the first of these papers, read before the Society on the 3rd and 10th of May, 1759, relates to undershot-wheels, and his definition of mechanical power and effect is so clear and concise, that it would be difficult to give a better explanation. He says:—

"The word *power*, as used in practical mechanics, I apprehend to signify the exertion of strength, gravitation, impulse or pressure, so as to produce motion; and by means of strength, gravitation, impulse or pressure, compounded with motion, to be capable of producing an *effect:* and that no effect is properly mechanical but what requires such a kind of power to produce it.

"The raising of a weight, relative to the height to which it can be raised in a certain time, is the most proper measure of power, or, in other words, if the weight raised be multiplied by the height to which it can be raised in a given time, the product is the measure of the power raising it, and consequently all those powers are equal whose products, made by such multiplication, are equal; for if a power can raise twice the weight to the same height, or the same weight to twice the height, in the same time that another power can, the first power is double the second; and if a power can raise half the weight to double the height, or double the weight to half the height that another can, those two powers are equal.

"But *note* that all this is to be understood in case of slow or equable motion of the body raised; for in quick, accelerated, or retarded motions, the *vis inertiæ* of the matter moved will make a variation.

"By the *vis inertiæ* of bodies is meant the force required to put them in motion when they are in a state of rest, or to accelerate them suddenly; a corresponding force must be exerted to stop or check a heavy body when it is in motion.

"In comparing the effects produced by water-wheels with the powers producing them,—or, in other words, to know what part of the original power is necessarily lost in the application, we must previously know how much of the power is spent in overcoming the friction of the machinery and the resistance of the air; also what is the real velocity of the water at the instant that it strikes the wheel, and the real quantity of water expended in a given time.

"From the velocity of the water, at the instant that it strikes the wheel being given, the height of head productive of such velocity can be deduced, from acknowledged and experimented principles of hydrostatics; so that by multiplying the quantity or weight of water really expended in a given time by the height of a head so obtained, which must be considered as the height from which that weight of water had descended in that given time, we shall have a product equal to the original power of the water, and clear of all uncertainty that would arise from the friction of the water, in passing small apertures, and from all doubts, arising from the different measures of spouting waters, assigned by different authors.

On the other hand, the sum of the weights raised by the action of this water, and of the weight required to overcome the friction and resistance of the machine, multiplied by the height to which the weight can be raised in the time given, the product will be equal to the effect of that power; and the proportion of the two products will be the proportion of the *power* to the *effect*: so that by loading the wheel with different weights successively, we shall be able to determine at what particular load and velocity of the wheel the effect is a maximum."

The nearer the effect obtained approaches to the power expended in obtaining it, the better and more perfect is the machine. Let the student bear in mind that the *effect* can never be greater than the *cause*. This is too often forgotten, and much money has been spent and time wasted in contriving machines to perform impossibilities.

The principles of mechanics are now so generally well understood that perhaps no person into whose hands this book may fall, will ever dream of inventing a machine which shall give a result equal to the motive power and overcome its own friction; yet the time has not long gone by when ingenious persons, reasoning on false premises, vainly flattered themselves that they might accomplish such things and devise some machine which should continue in *perpetual motion* without a maintaining power.

Mr. Smeaton goes on to show how the velocity of the water striking the wheel may be practically ascertained: first, by running the wheel unloaded in the water, and then by assisting it by means of counter-weights or cord wound round the axle, until the velocity of the wheel is identical with that of the water and the counter-weight, equal to friction and resistance of the air; for, if it were too little, the water would accelerate the wheel beyond the weight; and if too great, retard it; so that the water becomes a *regulator* of the wheel's motion, and the velocity of its circumference becomes a measure of the velocity of the water. The velocity thus determined, the virtual or effective head may be determined by the law of gravitation; and although, as Mr. Smeaton observed, the virtual head bears no certain or definite proportion to the actual head—as indeed has been shown in a foregoing part of this book—yet, when the aperture is greater, or the velocity of the water issuing therefrom is less, they approach nearer to a coincidence; and consequently, in large openings of mills and sluices, where great quantities of water are discharged from moderate heads, the actual head of water and the vertical head, determined from the velocity, will the more nearly agree, as experience confirms.

From the numerous experiments he had made on the undershot-wheel, Mr. Smeaton deduced the following rules, or, as he calls them, maxims:—

"1. That the virtual, or effective head, being the same, the effect will be nearly as the quantity of water expended.

"2. That the expense of water being the same, the effect will be nearly as the height of the virtual or effective head.

"3. That the quantity of water expended being the same, the effect is nearly as the square of its velocity.

"4. The aperture being the same, the effect will be nearly as the cube of the velocity of the water."

Upon comparing the several proportions between *power* and *effect*, remarked during the course of his experiments, Mr. Smeaton observes, the most general is that of 10 to 3, the extremes 10 to 3·2 and 10 to 2·8; but as it appears, that where the quantity of water, or the velocity thereof, that is, where the power is greatest, the second term of the ratio is greatest also, we may therefore well allow the proportion subsisting in large works as 3 to 1.

He also observes, that the proportions of velocity between the water and the wheel are contained in the limits of 3 to 1 and 2 to 1; but as the greater velocities approach the limit of 3 to 1, and the greater quantities of water that of 2 to 1, the best general proportion will be that of 5 to 2. He endeavoured to ascertain what is the ratio between the load such a wheel would carry at the maximum of effect, and what will totally stop it, and found that it would work steadily until that proportion was as 4 to 3; but when this limit was exceeded, the whole worked irregularly, and was liable to be stopped.

The principal aim, however, of a good millwright, is to make the wheel work to the greatest advantage; and the last-mentioned experiments were therefore more curious than useful; yet they are highly interesting, as they make the investigation complete, and anticipate a question which might very naturally be asked.

Mr. Smeaton mentions, that in his working model of an undershot water-wheel, the maximum load was equal to 9 lb. 6 oz., and that the wheel ceased moving with 12 lb. in the scale; to which, if the weight of the scale is added, nearly 10 oz., the proportion will be nearly as 3 to 4, between the load at the maximum and that by which the wheel is stopped: and he says,—

"It is somewhat remarkable, that though the velocity of the wheel, in relation to the water, turns out greater than one-third of the velocity of the water, yet the impulse of the

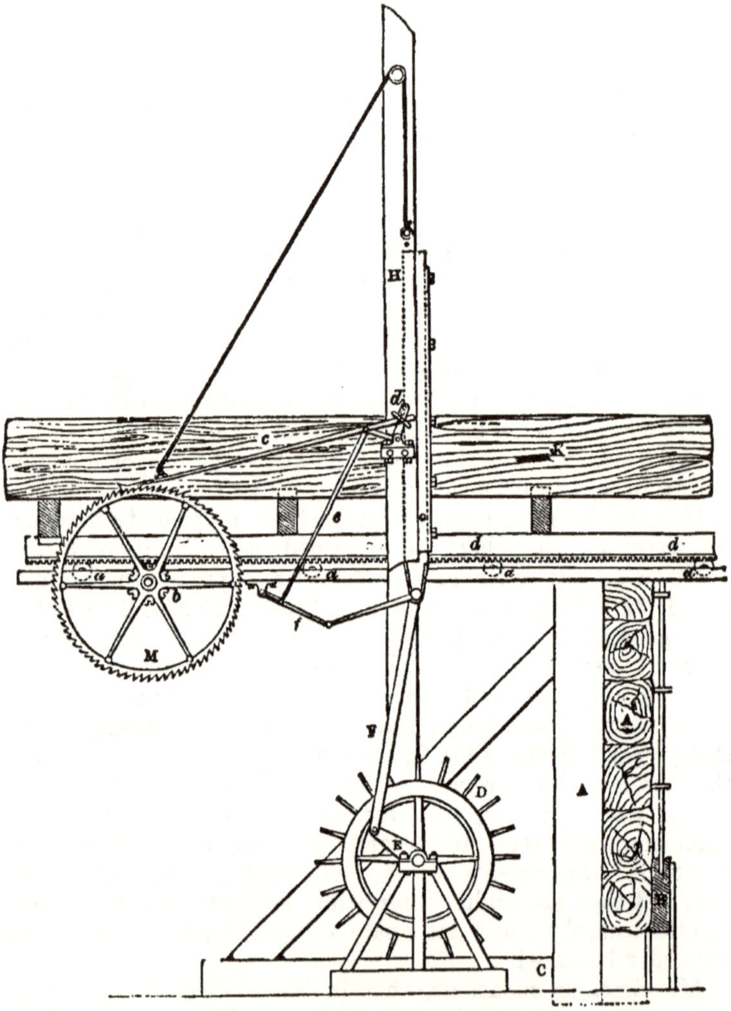

Fig. 26. Saw Mill constructed by Brunel.

water, in the case of a maximum, is more than double of what is assigned by theory; that is, instead of four-ninths of the column, it is nearly equal to the whole column."

It must be remembered, that in the present case the wheel

is not placed in an open river, where the natural current, after it has communicated its impulse to the float, has room on all sides to escape; but in a conduit or race, to which the float-board being adapted, the water cannot otherwise escape than by moving along with the wheel: and when a wheel works in this manner, as soon as the water strikes the float it receives a sudden check, and rises up against the float, like a wave against a fixed object; so that in the working model, when the sheet of water was not a quarter of an inch thick before it met the float, yet this sheet acted against the whole surface of a float three inches high; and, consequently, if the float were no higher than the thickness of the sheet of water, a great part of the force would be lost by the water dashing over the float.

Although in this country the value of water power, and the necessity to make the most of it, has gradually caused the undershot wheel to be abandoned, and the breast wheel to take its place, whereon the gravity or weight of the water acts instead of its impulse; yet there are many purposes to which the undershot wheel may be applied with advantage, and to none more than the sawing of timber, especially in new settlements, and in those localities where water power is abundant and mechanical skill is scarce, where labour is expensive and timber costs nothing. In such circumstances, a simple and efficient saw-mill is of great advantage, and it may be worked at once from the axle of the undershot water-wheel, working two saw frames, by means of cranked arms upon the ends of it; or, if the axle be made of iron, a crank may be formed in it to work a single saw frame, as shown in the annexed wood-cut, reduced from an engraving in the "Professional Papers of the Royal Engineers," vol. vi., in which will also be found a minute description, elaborately illustrated, of the saw mills and machinery for raising timber in Chatham Dockyard, erected by the late Sir Mark Isambard Brunel. (See Fig. 26.)

In this engraving A is the dam, frequently formed of square logs, resting against a standard secured by struts, provision being made to carry off the surplus water. B is the sluice, which, being raised to work the wheel, admits the water into the trough C; here it strikes the float-boards of the wheel D, which is generally made of small diameter, so that the velocity of the water may cause it to make as many revolutions as possible, consistent with the requisite power; the saws making as many strokes as the wheel makes revolutions. E, the crank on the wheel shaft, to which is adapted the connecting rod F, which is attached to the bottom of the saw-frame G. This slides up and down between the standards with an alternating motion, the strokes being double the length of the crank arm. K is the log to be cut; it is mounted on the frame *l*, which has a rack fixed in its under surface, and is supported by rollers *a a*.

The pinion *b* on the axis of the wheel M worked in the rack, and according as the wheel moves forward or backward, it works the frame, moving it towards or away from the saws. Motion is given to this wheel by the paul *c*, the other end of which is joined to one of the sides of the arm of the bent lever *d*. This lever is moved backward and forward by the rod *e*, which is joined to the bent rod *f* and this rod, or rather lever, is fitted upon an axle attached at one end to the frame of the wooden building, and at the other to the frame of the saw-mill. When it is requisite to reverse the motion, after the log is cut, the paul *c* is lifted clear of teeth of the ratchet-wheel M; and this wheel is turned in an opposite direction by hand.

A saw-mill on this principle was made by the late Mr. Rennie, of which he has given the following brief description: it appears to have been driven by the water of Leith. The diameter of the wheel was 4 feet 6 inches, and its width 3 feet 2 inches. The floats were 12 inches deep, and were inclined in direction 4 inches past the centre.

The crank had 11 inches radius; there was one on each end of the water-wheel axle; consequently the saw-frames driven by them had 22 inches stroke. The one frame had five saws and the other frame two saws; they made 13·6 strokes to cut one inch forward.

The fall of water was 3 feet 7 inches, and it was conducted in a sloping direction, in such a manner as to be nearly a tangent to it; the perpendicular height of all being 3 feet 7 inches, and the horizontal distance 8 feet.

The sluice was 3 feet 2 inches wide, and was raised to the surface of the water, namely 18 inches, at the head of the slope where it was placed. The sluice being raised to the water's surface, the wheel made fifty-eight turns per minute, working the seven saws. The timber cut was Norway fir; in the frame with five saws it was 8½ inches deep, and in the frame with two saws it was 6 inches deep, so that in one minute it cut 4·26 inches forward. It is to be observed, however, that in such small wheels there is a considerable loss of effect, when compared with wheels of larger diameter; and accordingly Mr. Rennie states the performance of another saw-mill, at Dartford, which, with the dimensions, he tabulates as follows:—

DARTFORD SAW-MILL.

1. Water-wheel 16 feet diameter.
2. Breadth 4 feet 6 inches, depth 1 foot 3 inches.
3. Fall 2 feet 3 inches, head 2 feet 9 inches.
4. Spur-wheel, or inner edge of shaft, 64 cogs.
5. Pinion on crank 18 cogs; thus the crank made 3·55 turns for one turn of the water-wheel.
6. Throw of crank 10 inches.
7. Slabbing-saw made 34 strokes for 4 inches advance of timber.
8. Deal-saw made 19 strokes to one inch advance.
9. Frame of saws 4 feet wide inside; saws 5 feet 3 inches long.
10. Ratchet-wheel 6 feet diameter.
11. Teeth of saws $\frac{3}{4}$ of an inch asunder.
12. Space taken out by saw $\frac{1}{8}$ of an inch.

EXPERIMENT.

1. Quantity of water, by floating wax balls, 2,461 cubic feet per minute.
2. Quantity, by measuring at the tail of the wheel, 3,297 cubic feet, medium 2,879.

But as the water was gathering in the pool, and as the river was extremely dirty, it is likely that the principal error may be in first experiments; and, therefore, instead of taking the medium quantity, we shall call it 3,000 cubic feet.

Sixteen saws at work in two frames, namely, two slabbing and fourteen deal saws; the water-wheel made twelve turns per minute. Thirty-four saws being put on, the wheel made only eight turns per minute.

CALCULATIONS.

The diameter of water-wheel being 16 feet, its circumference is $3{\cdot}1416 \times 16 = 50{\cdot}2656$; this multiplied by twelve revolutions $= 603{\cdot}18$ feet per minute, or 10·05 feet per second.

The head of water being 2 feet 9 inches, the velocity due

from this is 13·3; thus the velocity of the head of water is to that of the wheel as 13·3 to 10·05, or as 5 to 3·778; hence the effect produced will be nearly equal to twelve horse-power.

The reader will notice that the saws in the present instance are driven, not from the axle of the water-wheel, but from a second motion. The wheel is of larger diameter, and makes a less number of revolutions; for the requisite speed could not have been obtained from so low a head of water, and it is therefore gained by driving the pinion upon the second shaft, which works the saws, by the toothed wheel of greater size upon the water-wheel axle. The speed given to this wheel is somewhat greater than Mr. Smeaton advises, but this is done to render it more readily applicable to the saws, and the fall from the wheel clears it of back water.

The public are much indebted to the late Mr. George Rennie

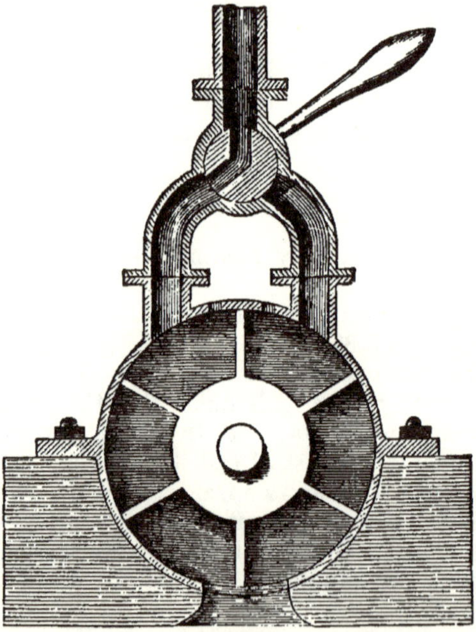

Fig. 27.

for the liberal manner in which he has given them these and other results of his father's experience, which will be found in the "Quarterly Papers on Engineering."

It is curious to remark, that the boards of the undershot

wheel still bear the name of *floats*, although they are no longer turned by the current of the river; while the cavities or receptacles for the water on the face of the overshot wheel are called buckets, although they bear no resemblance to a bucket, otherwise than by holding water. In these names we trace their origin from the ancient irrigating-wheel, turned by floats in the stream, and lifting the water in wooden buckets or earthen pitchers fastened upon the rim of the wheel.

A very simple and neat method of employing the impulsive force of water, when a small stream descends from a great height, is shown by a working model in the Museum of Practical Geology, London. A jet of water issuing from a pipe is made to strike the vanes of a small wheel, enclosed in a case, and to cause its rapid revolution. The wheel may be stopped, or its motion reversed, by turning a stop-cock or cylinder. The whole machine may be made of brass, or other metal, in a very compact form; and in a hilly country might be usefully employed at a farm in turning agricultural machines, to save manual labour. Its small size and cost makes it well suited for such purposes. (See Fig. 27.) The model at the Museum in Jermyn Street is adapted to machinery for winding up ore from a mine.

CHAPTER IX.

OVERSHOT WHEELS, GRAVITY OF WATER EMPLOYED INSTEAD OF IMPULSE.

It was not difficult to imagine that if a small stream of water descending from a hill side were directed into the mouths of the earthen vessels or wooden buckets of the wheels used for irrigation, the vessels so loaded would descend and the wheels revolve, so that rotary motion and mechanical power would be gained; the buckets emptying themselves at the lowest point, as they had before been emptied at the highest; the wheel turning in the opposite direction, because the weight, or gravity of the water was now the moving power of this OVERSHOT WHEEL.

In the undershot wheel the impulse of the water striking the floats drives the wheels; in the overshot wheel the weight of the water flowing into the buckets turns the wheel, and all impulse must be avoided; the water must flow with the same velocity as the wheel, or just so much in excess as will prevent the buckets from striking the water as they present themselves to be filled. Experience soon showed that the earthen jar or the suspended bucket were cumbrous and inconvenient, and as larger and more powerful wheels were applied to more copious streams, a series of simple wooden troughs formed across the face of the wheel were found to answer the purpose better. When the supply of water was ample and the wheels large, it was found that to fill these troughs well and regularly the stream should be made nearly as broad as the wheel, and shallow in proportion to its width. The wheel was then formed by placing two sets of *arms*, at a sufficient distance apart, upon the *axle*, and fixing to their ends *segments* of wood to form the circle; upon these segments across *the face* of the wheel, and equal to, or somewhat exceeding in length the width of the stream or sheet of water, were nailed the *sole-boards;* on the end of these boards, and at right angles to them, so as to form a projecting rim or ledge on each side of the wheel's face, was fixed the *shrouding*, formed of stout plank generally from 12 to 18 inches broad; and between these shroudings, across the face of the wheel, were placed the *buckets*, made of lighter planking, and having their ends let into the shrouding, by which the ends were closed. The edge of the bucket-board meeting the sole-plank formed two sides of a triangular trough, the third being open to receive the discharge of water. Subsequently the bucket was made in two boards, one called the front, and the other the bottom of the bucket, the latter taking off the angle and making the section of the bucket, or form of the trough, that of a trapezium, which form it long retained, until the buckets-of-water wheels were made of iron plate.

Since water-wheels have been made wholly of iron, and chiefly of wrought-iron, the form of the bucket has been either a part of a circle, a cycloid, an epicycloid, or an Archimedian spiral. These forms are noticed in a subsequent page in connection with breast wheels. Great pains are now taken by the best makers of water-wheels to form and adapt the curve of the buckets so that they may readily fill with water, retain their load as long as possible, and discharge it with facility when it has ceased to be useful.

Mr. Smeaton's paper on undershot wheels, before quoted, was followed by another on overshot wheels, read before the

Royal Society, May 24, 1759. These papers were long considered as the text from which millwrights should deduce their rules of practice; and even now they well deserve the careful study of those who engage in the construction of such machines, together with the later experiments of Rennie, Morin, and Poncelet.

Mr. Smeaton had the merit of proving and demonstrating the advantage and the difference of effect resulting from employing the weight instead of the impulse of a volume of water descending from a given height.

"In reasoning without experiment, one might be led to imagine that, however different the mode of application is, yet that wherever the same quantity of water descends through the same perpendicular space the natural effective power would be equal; supposing the machinery free from friction, equally calculated to receive the full effect of the power, and to make the most of it: for if we suppose the height of a column of water to be 30 inches and resting upon a base or aperture of 1 inch square, every cubic inch of water that departs therefrom will acquire the same velocity or *momentum*, from the uniform pressure of 30 inches above it, that 1 cubic inch let fall from the top will acquire in falling down to the level of the aperture; one would therefore suppose that a cubic inch of water let fall through a space of 30 inches, and then impinging upon another body, would be capable of producing an equal effect by collision, as if the same cubic inch had descended through the same space with a slower motion, and produced its effects gradually; for in both cases gravity acts upon an equal quantity of matter, through an equal space; and, consequently, that whatever was the ratio, between power and effect in undershot wheels, the same would obtain in overshot, and indeed in all others; yet, however conclusive this reasoning may seem, it appears upon trial, that the effect of the gravity of descending bodies is very different from the effect of the stroke of such as are *non-elastic*, though generated by an equal mechanical power."

Gravity, it is true, acts for a longer space of time upon the body that descends slowly, than upon one that falls quickly: but this cannot occasion the difference in the effect; for an elastic body falling through the same space in the same time will, by collision upon another elastic body, rebound nearly to the height from which it fell: or, by communicating its motion, cause an equal one to ascend to the same height.

The observations and deductions which Mr. Smeaton made from his experiments were as follows—

First.—As to the ratio between the power and effect of overshot-wheels.

The effective power of water must be reckoned upon the whole descent; because it must be raised to that height, in order to be in a condition for producing the same effect a second time.

The ratio between the powers so estimated, and the effect at the maximum as deduced from the several sets of experiments, is shown to range from 10 to 7·6 to that of 10 to 5·2; that is nearly from 4 to 3 and from 4 to 2. In these experiments, where the heads of water and quantities expended are least, the proportion is nearly as 4 to 3; but where the heads and quantities are greatest, it approaches nearer to that of 4 to 2, and by a medium of the whole the ratio is that of 3 to 2 nearly. We have seen before, in our observations upon the effects of undershot wheels, that the general ratio of the power to the effect when greatest was 3 to 1; the effect, therefore, of overshot wheels, under the same circumstances of quantity and fall, is, at a medium, double to that of the undershot.

Second.—As to the proper height of the wheel in proportion to the whole descent.

It has been observed, that the effect of the same quantity of water descending through the same space is double, when acting by its gravity upon an overshot wheel, to what the same produces when acting by its impulse upon an undershot. Therefore the whole height at the fall should be made available, because, when the water is laid upon the top of the wheel, it is upon the gravity, and not the impulse, that the effect depends. A sufficient fall, however, must be given to lay on the water with a velocity *somewhat greater* than that of the circumference of the wheel, otherwise the wheel will not only be retarded by the buckets striking the water, but a part of it will be dashed over and lost, while the buckets will not be so well filled; but no greater velocity should be given than is sufficient to accomplish these objects, as it would be power wasted.

Third.—As to the best velocity of the wheel's circumference in order to produce the greatest effect.

If a heavy body fall fairly from the top to the bottom of the descent, it will take a certain time in falling, but during the fall no mechanical effect is produced; for in this case the whole action of gravity is spent in giving the body a certain velocity; but if this body in falling be made to act upon something else, so as to produce a mechanical effect, the falling body will be retarded, because a part of the action of

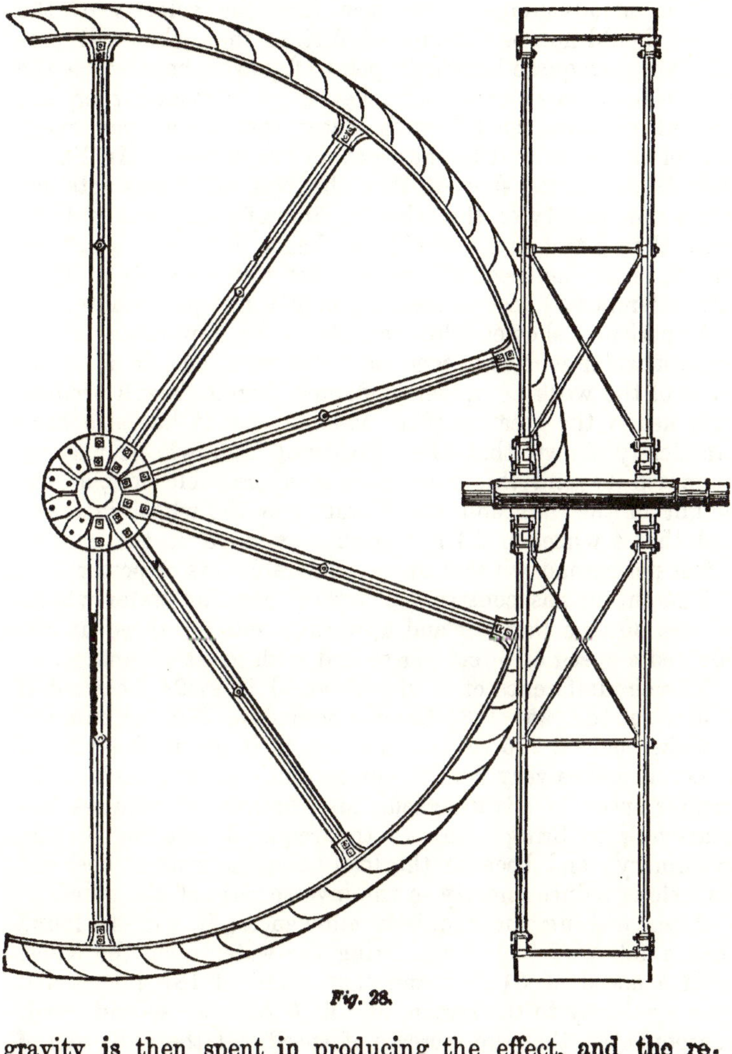

Fig. 28.

gravity is then spent in producing the effect, and the remainder only in giving motion to the falling body; and therefore the slower a body descends, the greater will be the action of gravity applicable to produce a mechanical effect.

If an overshot wheel had no friction, or other resistance, the greatest velocity it could attain would be half a revolution in the same time that a heavy body laid upon the top of it would take to fall through its diameter, but no mechanical effect could be derived from the wheel.

It is an advantage in practice that the velocity of the wheel should not be diminished further than what will procure some adequate benefit in point of power, because, as the motion becomes slower, the buckets must be made larger, and the wheel being loaded with water, the stress upon every part of the work will be increased in proportion. Mr. Smeaton's experiments showed that the best effect was obtained when the velocity of the wheel's circumference was a little more than 3 feet in a second; and hence, it became a general rule to make the speed of the overshot water-wheels at their circumference 3½ feet per second, or 210 feet per minute.

Experience showed this velocity to be applicable to the highest water-wheels as well as the lowest, and if all other parts of the work be properly adapted thereto, it will produce very nearly the greatest effect possible; but it has also been practically shown that the velocity of high wheels may be increased beyond this rate without appreciable loss, as the height of the fall and the diameter of the wheel increase, and that a wheel of 24 feet high may move at the rate of 6 feet per second without any considerable loss of power.

The author has constructed several overshot water-wheels of iron 30 feet diameter and upwards; and for these he has adopted a speed of 6 feet per second with great advantage.

The circumference of a 30 feet wheel is 92·24 feet, and if it move at the rate of 3½ feet per second, or 210 per minute, it makes only 2·275 revolutions, but if it go at the rate of 6 feet, it makes very nearly four revolutions per minute. The greater speed is advantageous, also, because it requires less gear-work to bring it up to the required rate for driving machinery; and because the load upon the water-wheel and its axle is reduced nearly in the inverse ratio of the speeds.

After making the requisite allowances, it will be found that a fall of 7 inches will bring the water upon the wheel with a speed of 3½ feet, and that a fall of 18¾ inches will give a velocity to the stream of full 6 feet per second; and, consequently, that the increase of speed and the reduction of load upon the wheel, as before stated, are gained at the expense of 1 foot of fall, making a difference of effect between a wheel of 31 feet diameter, at the slower speed, and a wheel of 30 feet diameter at the greater, in the ratio of 44·6 to 43·0; but the steadiness and regularity of motion derived from the momentum of the quicker wheel render this difference practically inappreciable as compared with the gain. By the foregoing comparison it will be seen how far the rule laid down by Mr. Smeaton may in practice be departed from as the diameter of the wheel increases.

The diameter, however, has its limits, and a water-wheel of great height is costly, cumbrous, and slow. A wheel was erected in South Wales by Messrs. Crawshay, at the Cyfarthfa Iron Works, near Merthyr Tydvil, 50 feet in diameter and 6 feet wide: it had 156 buckets, and made 2½ revolutions per minute. This wheel, erected in the year 1800, was used to blow the furnaces. It was for some time the largest and most powerful water-wheel in use; but it has since been considerably surpassed. Mr. John Taylor applied a fall of water in the mining district, near Tavistock, 526 feet in height, to give motion to seventeen water-wheels; eight of which were employed in pumping water from a depth of nearly 200 fathoms. The diameter of the largest of these wheels was 51 feet, with a width of 10 feet in the clear across the face; whilst the smallest of the eight wheels was 32 feet in diameter; the others were of intermediate size. Four other wheels gave motion to machinery for drawing up the ores to the surface; and the remaining five were employed to drive mills for crushing and stamping the ores.

The performance of the largest wheel was as follows:—diameter, 51 feet; breadth, within the shrouding, 10 feet; the water poured into its buckets was at the rate of 5,632 gallons per minute, which, at 10 lbs. per gallon, would be 56,320 lbs. weight descending 51 feet = 2,872,320 lbs. descending 1 foot or to 87 horse-power.

The wheels, when so supplied, made five revolutions per minute, and worked six pumps of the diameters and depths annexed. The length of stroke in each pump was 6 feet, and the effective stroke or motion in the pumps to raise water was at the rate of 30 feet per minute.

Lifts.	Fathoms.	Feet.	Diameter.	Weight, lb.
1	43	1⅗	13½ in.	16,136
3	112	3½	13 „	38,922
1	25	3	14 „	10,225
1	6	5	9 „	1,132
				66,415

The total weight of the water raised by the pumps was

66,415 lbs.; as the rate was 30 feet per minute, 1,992,450 lbs. were raised 1 foot per minute; that is, the force actually realised was equal to 60·36 horse-power.

The useful effect, or work done, being at the rate of 69·4 per cent. of the power expended, the remaining 30·6 per cent. was lost, owing to the friction of the pump-work, the resistance to the motion of the water through the pump, and other retarding causes; this, however, is good duty for such a wheel so applied.

A wheel larger in diameter, but not so powerful, has been made by Messrs. Donkin and Co., of London—namely, 76½ feet. The width of face is 2 feet, the number of buckets 160, and the power stated to be 30 horse. It is all of iron, and principally of wrought iron. It was sent to Italy, where it has been erected, but the effect has not been ascertained.*

Attempts have been made to employ a high fall of water by placing one wheel above another; this was tried many years ago at Aberdare, in South Wales, where two wheels, each 40 feet in diameter, were so placed, like the figure of 8, and were connected by teeth on their respective rims—the lower wheel receiving the water after it left the upper one, and revolving in the opposite or reverse way. The result was not satisfactory; but in another case, a drawing of which lies before the writer, wherein Messrs. Charles Wood and Brothers, of Macclesfield, had two overshot water-wheels each of 26 feet in diameter, and 6 feet wide, placed over each other, they succeeded by a somewhat different arrangement of the toothed-wheel work. The two wheels were not connected immediately with each other, but by means of *pinions*, which worked into teeth upon the rims of the two water-wheels, causing them both to revolve in the *same* direction, so that the water, on leaving the buckets of the upper wheel, was more easily and readily received by the buckets of the lower wheel.

In either of these cases, however, the employment of the turbine, or the *pressure engine*, which will be described hereafter, would have been much less costly and more effective. The like may be said of all the contrivances to substitute endless chains with buckets applied to high falls instead of water-wheels.

* A still larger overshot water-wheel has been constructed in Ireland by Messrs. John and Robert Mallet, of Dublin, and is employed driving a large paper-mill within a few miles of that city. It is 80 feet in diameter, and 8 feet wide on the breast. The shrouding buckets, centres, shafts, &c., are of iron, and the arms of oak timber.

Where the quantity of water is large and variable, and the fall such as may be termed an intermediate height, but varying also with the supply, it is found advantageous not to lay the water upon the top of the wheel, so that it may work overshot, but to make the diameter of the wheel greater than the mean height of the fall, and to lay the water, as it were, "*on the shoulder*" of the wheel, or at 45 degrees from the perpendicular; that is, half-way between the horizontal line and the perpendicular, or, as millwrights say, "at nine o'clock." Very little mechanical effect is produced in the upper eighth of the circle as compared with the next quarter, on which the descent of the water is nearly perpendicular, and when the wheel is fitted with toothed segments at or near its circumference, acting on a pinion placed on a level with the axle, the weight of the water is brought to bear at once upon the pinion teeth, the stress is taken off the arms of the wheel, and the axle becomes, as it were, merely a pivot on which the wheel turns. By this arrangement, the late Messrs. Hughes and Wren, of Manchester, were enabled to make the arms of their wheels of simple tension rods of bar-iron, by which the rim of the wheel was tied and braced to the centre, a plan which, with some modifications and improvements, is still in use, and sometimes the segments have interior teeth, which renders the wheel-work more compact.

In the best constructed wheels, the water is laid on in a thin sheet of no greater depth than will give it a somewhat greater velocity than that of the wheel, the difference being just sufficient to pour into the succeeding buckets the proper supply of water. The buckets should be so capacious that they need not be full when the wheel carries its maximum load, in order that no water may be wasted, and that they may retain the water in them till the last moment that its weight on the wheel is effective, and yet empty themselves as soon as it ceases to be so. It is also expedient in practice to make the width of the sheet of water less than that of the wheels: if the wheel be broad on the face, the stream may be 4 inches shorter than the length of the buckets: the air escaping at the ends is thus prevented from blowing out the water; and all these precautions, though small in themselves, tend to produce smoothness, regularity, and increased effect in the working of the machinery.

There is, however, one mode of using water-power—acting by its gravity—in buckets upon a chain, much employed in South Wales, which is found very useful for raising ore from the pits. An endless chain is passed over a wheel of 16 feet in diameter, placed between two shafts. The chain passing down each shaft, and through an opening at the bottom between the two: two large buckets, or rather shallow tubs, of wrought iron, are fixed upon the chain, so that the suspension is by the centre of the tubs, and they are so placed that when one tub is at the top of its shaft, the other is at the bottom of its shaft. Each tub or bucket is covered by a strong platform, which fills and closes the pit's mouth when hoisted up, and carries the small waggon or tram containing the ore upon it; and each is also fitted with a valve at the bottom to discharge the water. A branched pipe, communicating with an elevated reservoir, is laid to the mouths of the shafts, and fitted with stop-cocks or valves. The tub at the surface being filled with water, over-balances the empty tub at the bottom, and raises it, with its tram load of ore, to the top. When the full bucket has descended the shaft, the valve is opened and the water discharged; the other being filled in like manner, descends, and thus alternately each raises the other with its load of oar. The water finds its way out of the mine by a drift or adit into the valley; the long loop or bight of slack chain below the buckets, and hanging to the centre of each, equalises the weight of chain at all times; and a brake applied to the large wheel regulates the speed of the descending bucket. In some places, the two buckets work in one shaft of an oblong form; the diameter of the wheel is reduced to 7 feet; it is fitted with toothed segments, working into a pinion, fixed upon a second axle, on which the brake wheel is placed, in order to gain the requisite power to control the descending weight. Drawings of both these plans lie before the writer, but the principle and construction are so simple that a description will probably suffice. It may be proper to mention that the buckets generally work in guides, that the discharging valves are opened by striking upon a point or projecting spike at the bottom of the shaft, and that upon the platforms which cover the buckets there is a portion of the rail or tramway laid to match with the lines of way at the top and bottom of the shaft, so that the tram or carriage may run from the platform to its destination.

CHAPTER X.

BREAST WHEELS; M. PONCELET'S WHEEL. LARGE STREAMS AND LOW FALLS.

AT the time when Mr. Smeaton wrote the papers on undershot and overshot water-wheels, before referred to, the BREAST-WHEEL was but little known and imperfectly understood; for it seems then to have been a kind of compromise between the two, in which it was attempted to combine the action of impulse and gravity. It had become evident that the impulse of a column of water did not produce the same effect as its weight; and the experiments of Mr. Smeaton showed what the difference was. Having investigated the other two modes of applying water-power, he contents himself with saying, "We might naturally proceed to examine the effect where the impulse and weight are combined, as in the several kinds of breast-wheels, &c.; but the application of the same principles in these mixed cases will be easy, and reduce what I have to say on this head into a narrow compass; for all kinds of wheels where the water cannot descend through a given space, unless the wheel moves therewith, are to be considered of the nature of an overshot wheel, according to the perpendicular height that the water descends from; and all those that receive the impulse or shock of the water, whether in a horizontal, perpendicular, or oblique direction, are to be considered as undershots; and, therefore, a wheel, which the water strikes at a certain point below the surface of the head, and, after that, descends in the arc of a circle, pressing, by its gravity, upon the wheel; the effect of such a wheel will be equal to the effect of an undershot, whose head is equal to the difference of level between the surface of water in the reservoir and the point where it strikes the wheel, added to that of an overshot, whose height is equal to the difference of level between the point where it strikes the wheel and the level of the tail water."

It is here supposed that the wheel receives the shock of the water at right angles to its radii, and that the velocity of its circumference is properly adapted to receive the utmost advantage of both these powers; otherwise a reduction must be made on that account.

Fig. 29.
Breast Wheel.

Here Mr. Smeaton leaves the subject; his remarks having suggested considerable improvements on the common practice of his day, and dispelled many popular errors. The principles he developed, and the rules he deduced from them, were productive of great practical benefit; and his papers may still be studied with much advantage.

The late Mr. Rennie may be said to have continued these researches; and, in 1784, he erected a large breast-wheel, worked by the weight of the water alone. To this wheel he first applied *the sliding hatch*, over which the water flowed upon the wheel, instead of issuing under the hatch or sluice, as formerly; and by this means the whole height of the fall is realised.

In this case, the sluice or hatch is to be drawn up, in order to stop the flow of water, which runs over it; whereas, before this time, it was let down to shut off the water running out below.

The water ran over the top of the sliding hatch, as over a dam, in a thin sheet, and was laid upon the *breast* of the wheel, or below the level of the axle, where it acted, by its gravity, upon the float-boards; being held up to the breast by masonry cut to a circular arc, drawn from the centre, fitting the wheels' circumference, and confining the water on all sides, so that none of it could escape without doing its part in the work; for the float-boards filled the channel, and there were sole-boards upon the rim of the wheel. Here let it be observed, that these sole-boards, which form an obtuse angle with the floats, are notched upon the oak starts to which the float-boards are fixed; but they do not completely close the wheel rim; a long narrow slit or opening is left across the face of the water-wheel, between the starts, permitting the air to escape, and the water freely to enter as each bucket or float presents itself to receive its load; and should the river be flooded, and rise against the wheel, so that it "runs in backwater," the ready admission of air from above prevents a partial vacuum from being formed between the floats, which would tend to retard the escape of the water, and thus reduce the useful effects. Although in this wheel, which is without "shrouding" or sides, there is but little cavity or depth to cause such retention, yet in large wheels, with well-buckets and deep shrouding, the loss of power would be important. It has always been the author's practice, even in over-shot wheels, to make provision for ventilating the buckets, and prevent "the suction," as workmen say, from the tail water.

As, however, there is sometimes a little water lost by this means, and the orifice so left is more effectual when made larger, especially when buckets of much capacity are used instead of floats, as they now frequently are, in breast-wheels, an ingenious method of ventilating the buckets has been proposed by Mr. Fairbairn, which is well worth the little extra cost it may occasion in the first outlay of capital required for the erection of a large wheel.

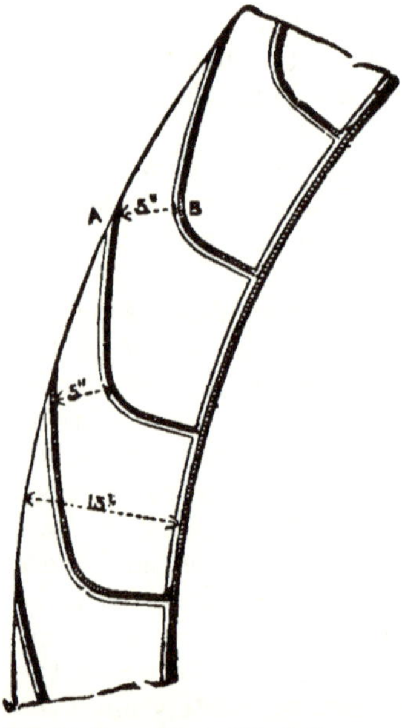

Fig. 30. Proportions of Buckets.

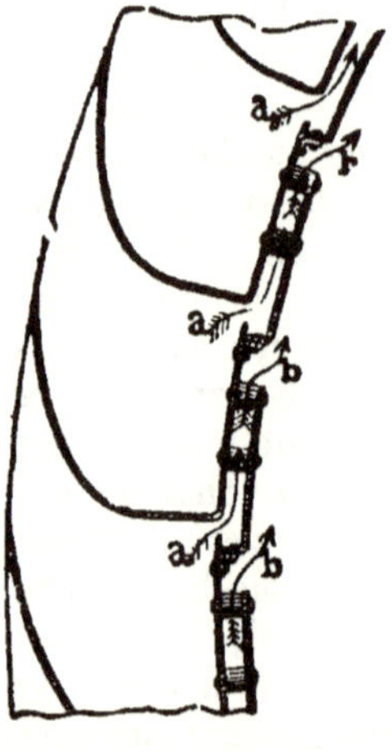

Fig. 31. Ventilating Buckets.

In order to lay the water upon these wheels with the velocity required, and, at the same time, to keep the stream in a thin sheet, Mr. Rennie next adopted a double hatch, or sluice—adding, as it were, another hatch, fixed above the first; so that, by an adjustment of the two, a narrow opening might be left between them, capable of nice regulation to every variation of power required for the quantity of work to be done in the mill. The upper hatch dipped just so much below the water's surface as to give the needful head above the orifice to produce a jet flowing with the desired speed; and the under hatch being raised or lowered by racks and pinions, the thickness of the stream delivered upon the wheel was diminished or increased at pleasure, while it continued to drive the mill at an uniform rate, by varying the load or quantity of the water.

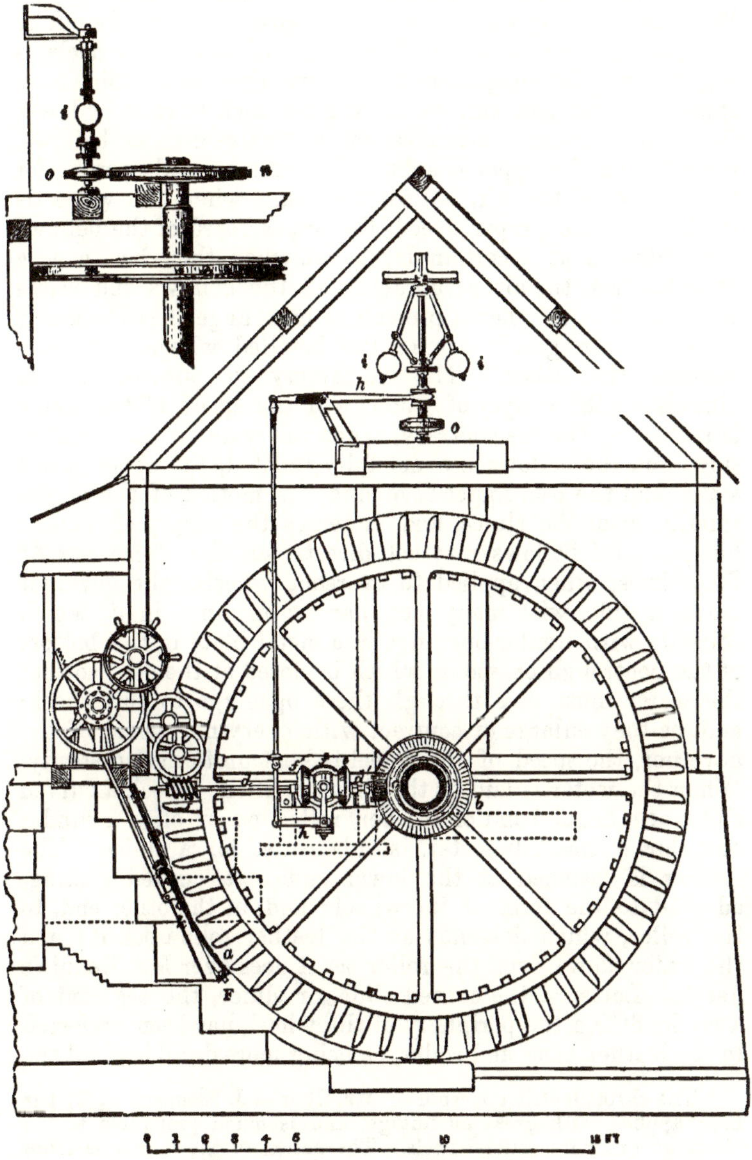

Fig. 32. Water-wheel at Waltham Abbey.

The next improvement was the application of the governor,

or regulator, to adjust the supply of water to the wheel without the attention of the miller. The double revolving pendulum, adopted as a regulator for the steam-engine by Mr. Watt, is generally used; and as the governor-balls fly out or collapse, the one or other of the pair of bevilled wheels is engaged and disengaged, so as to turn the pinion spindle in opposite directions, and so moving the rack to raise or lower the sluice. Some mechanics use a strong circular bellows, weighted on the upper boards, and having a small opening for the escape of the air, so adjusted that when the wheel is working at the proper speed, the upper board of the bellows is maintained at a certain level, but when the wheel goes a little too fast, the air accumulates in the bellows and raises the upper board, when a projecting iron finger fixed upon it "strikes into gear" one of the bevilled wheels, and disengages the other. The machinery so put in motion diminishes the supply of water, and the speed of the whole is reduced. On the contrary, when the water-wheel goes too slow, and the bellows are not fully inflated, the upper board sinks, and the iron finger reversing the motion of the pinion spindle opens the sluice, and increases the supply of water; but powerful bellows may act directly on the sluice itself.* Sometimes, when applied to overshot wheels, the governor moves a slide, with many long narrow openings in it, which coincide with similar openings in a metal plate in the bottom of the pentrough or spout, which is closed at the end, so that the water must pass through these openings to the wheel; and, as they enlarge or contract with every movement of the governor, the speed of the wheel is kept uniform and steady. When the water is laid on the shoulder of the wheel, a curved slide, or a broad band of leather, rolled on or off a cylinder or roller, of small diameter, is substituted for a sluice. The leather is fastened at the lower end to a curved grating, adapted to the form of the wheel; and, at the other end, to the roller, which descends as the leather rolls upon it; and the water flows over the roller as it does over Mr. Rennie's hatch. Latterly, the curved slide, or sluice, the segment of a circle, fitting the periphery of the wheel, has been preferred to the leather band and roller, as being more durable; and the

* The chronometric governor of Mr. Charles J. Siemens, C.E., has been applied with great advantage, and is much preferable to the Watt governor for water-wheels. The action of the former is more sensitive and prompt on sudden alterations of water supply, or of resistance in waste. When the Watt governor is employed, Porter's modification is the best form to give sensibility.

sluice is now better fitted, since improved tools have been introduced in the manufacture of machinery.

There are several other modes of applying governors; but these will, perhaps, be considered sufficient.

Breast-wheels are often made wide across the face, and of considerable power. One of the widest was erected many years ago, by Messrs. Strutt, at their Belper cotton-mills. It was 40 feet wide, but only 12 feet 6 inches diameter; the shaft was made hollow, like a cask, and so large as to form the body of the wheel, the float-boards being fixed immediately upon it. This hollow axis was 48 feet long, composed of thirty-two staves, 6 inches thick, bound together by iron hoops; the diameter in the middle was 7 feet 2 inches, tapering to 5 feet at one end, and 6 feet at the other, upon which a toothed wheel, 14 feet in diameter, was placed. The ends of this barrel were made solid for 3 feet at the small end, and 4 feet at the larger, to receive the gudgeons or pivots on which it turned. The float-boards, twenty-four in number, did not come in immediate contact with the barrel, but a space of 2 inches was left for the air to escape. The length of the float-board was divided into ten parts, and these were supported by rings fixed upon the hollow axis, placed 4 feet apart. The divisions of the float-boards were not placed in a line with each other, but every one in advance of the next, by equal steps, so that the water might not strike the whole length of 43 feet at once, but act in a rapid succession of intervals, and thus make the stroke imperceptible.

A pair of very powerful wheels were erected many years ago at the Katrine Works, in Ayrshire, by Messrs. Fairbairn and Lillie. It was originally intended to have four of these wheels, and provision was made for receiving them, should the factory be extended and require more power; but at present two wheels are sufficiently powerful, except in very dry seasons, to turn the whole work of the mills.

These wheels are each 50 feet in diameter; 10 feet 6 inches wide inside the buckets; and 15 inches deep in the shroud; they are fitted with internal toothed segments forming a circle of 48 feet 6 inches diameter; the teeth are 3¼ inches pitch, and the breadth is 15 inches. The fall of water conveyed from the river Ayr is 48 feet, and the power of the two wheels 240 horses.

When iron is used to make water-wheels, as is now very generally the practice, curved buckets and shrouding are preferred to deal float-boards and oak starts.

In designing these iron wheels, care should be taken that,

while attempting to form a bucket which shall carry the water as long as it can act beneficially, the openings or mouths of the buckets shall not be so contracted as to prevent the free admission of water into them, or its free discharge at the bottom of the wheel.

Mr. Fairbairn from his experience states that, in the construction of wheels for high falls, the best proportion of the opening of the bucket is proved to be nearly as 5 to 24; that is, the contents of the bucket being 24 cubic feet, the area of the opening, or entrance for the water, would be 5 square feet.

In breast-wheels, which receive the water at a height of ten to twelve degrees above the horizontal centre, the ratio should be nearly as 8 to 24, or as 1 to 3. With these proportions, the depth of the shrouding is assumed to be about three times the width of the opening, or three times the distance from the lip to the back of the bucket, the opening being 5 inches, and the depth of the shroud 15 inches.

These proportions are shown in Fig. 30.

For lower falls, or those wheels which receive the water below the horizontal centre, a larger opening becomes necessary for the reception of a larger body of water, and its final discharge.

In the construction of water-wheels it is requisite, in order to obtain the maximum effect, to have the opening of the bucket sufficiently large to allow an easy entrance and an equally free escape for the water, as its retention in the bucket must evidently be injurious when carried beyond the vertical centre, or perpendicular line below the wheel's axis.

Another mode of applying the water to wheels worked by low falls has been introduced by M. Poncelet, whose works on hydraulics, and especially on water-wheels, are well known; his mode is fully elucidated in his "Mémoire sur les Roues Hydrauliques à Aubes-Courbes, mues par-dessous." ("Report on Water-wheels with Curved Buckets, worked from below.")

M. Poncelet's method is especially well suited for falls under eight feet in height, and where the quantity of water is large, as by this mode of working the wheel from below, it is not subjected to a load of water, but receives it, as it were, under pressure. This wheel consequently can be made very light, although of great power, with arms of wrought-iron, like the paddle-wheels of a steam packet; and as it can be driven at a greater rate than the breast-wheel, a larger quantity of water

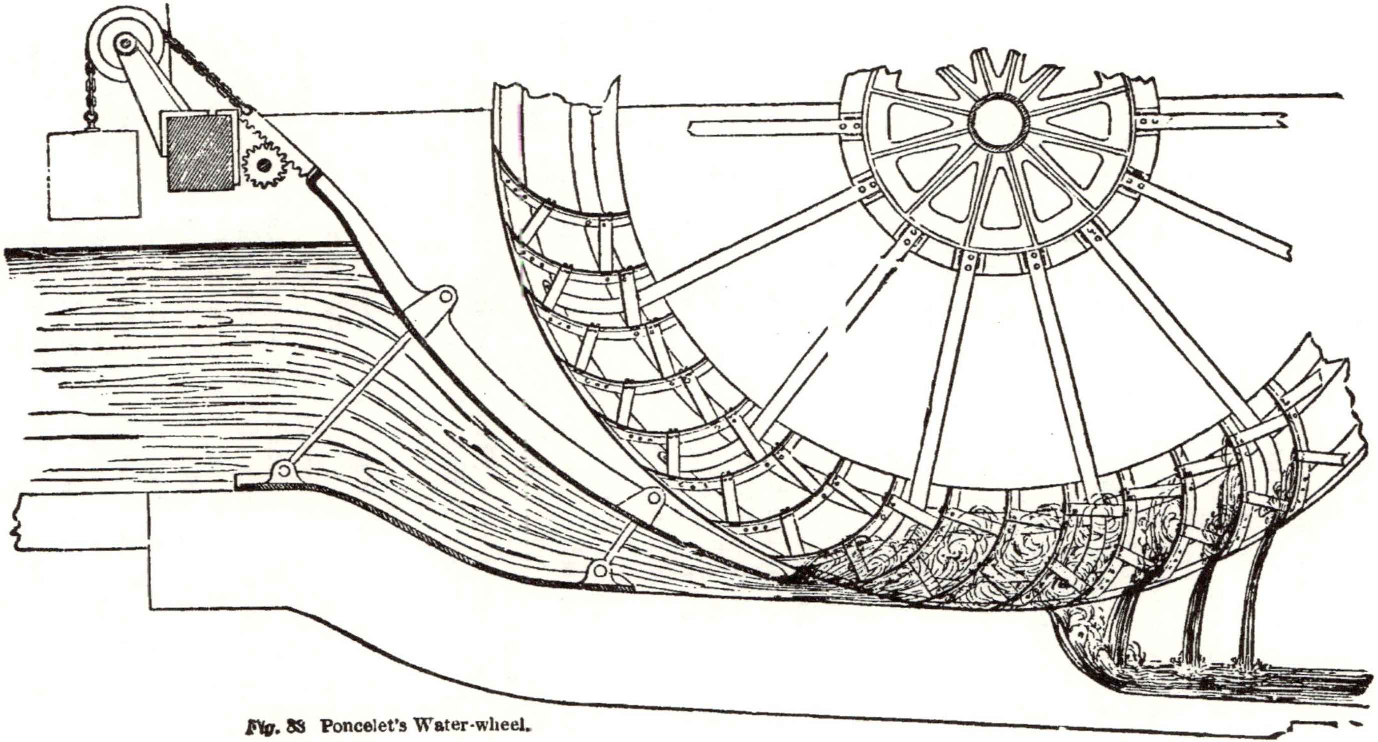

Fig. 83 Poncelet's Water-wheel.

may be brought to bear upon a narrower wheel. It is fitted with curved buckets, deeper than those of the breast-wheel, and without backs or sole-boards, so that within the rim of the wheel they are altogether open, and as the air can offer no impediment to the water's entrance or exit, the buckets are more numerous and their mouths narrower; one great object in this design being to make the action of the water continuous; avoiding the shock experienced in the undershot-wheel; the water enters them nearly at the lowest point of the wheel, and at a tangent to it, issuing from beneath a curved sluice, which is opened by being drawn upwards by racks and pinions. This sluice is nearly in contact with the wheel, so that the thin broad stream of water acts directly upon the buckets with all the pressure due to its head; and as they present themselves in rapid succession, this pressure is almost constant.

The sluice is placed in an inclined position, leaning as it were against the water, and is held down by radius bars extending from the back of it to the masonry below; these bars, being jointed at both ends, serve not only to retain the sluice in its proper place, but to guide it as it opens or shuts down upon the sill. The sluice is made of cast-iron, bolted together with flanged joints, planed to fit close, and form one large plate of equal breadth with the wheel; and at the sides it is kept tight by coming close to the stonework, or to cast-iron sides, against which a strong leather packing, secured to the sides of the sluice, is pressed by the force of the water.

The woodcut (Fig. 33) is taken from a drawing exhibited at the Institution of Civil Engineers, and represents a water-wheel on M. Poncelet's system, erected by M. De Bergue, who stated that this wheel was 16 feet 8 inches in diameter, and 30 feet wide; that when driven by a fall of water 6 feet 6 inches high, it yielded 120,000 cubic feet per minute; that it gave about 180 horse-power when the circumference moved at the rate of 11 to 12 feet per second: that a breast-wheel, worked by the gravity of the water in the usual way, and travelling at the ordinary rate, must have been 90 feet wide to produce the same effect; and that the speed of the wheel was about fifty-five per cent. of that of the water.

Referring to M. Poncelet's "Mémoires," before mentioned, in the second part, "Sur des Experiences en Grand," after having given the most elaborate details of his experiments, and tabulated them with all the method in which the French engineers excel, he says, under the head of *Calcul de la Vitesse et de la Force de la Roue*,—"With the various data already given, we shall be in a position completely to determine in a manner sufficiently exact for practice, all the dimensions of the wheel; to calculate its most advantageous velocity, the power it will transmit to the machinery when working at the maximum of effect, and the effect it is capable of exerting tangentially to its circumference.

"With respect to the most advantageous speed of the wheel, it may be admitted, in accordance with the results of the experiments, that it is limited to between 50 and 60 per cent. of *the velocity due theoretically to the water level above the centre of the orifice*, the last number having relation to low falls and large openings, the other to high falls and small openings of the sluice; so that for the medium falls of 1·3 mètres (about 4 feet 3 inches) above the sill of the aperture, with medium openings of 20 centimètres (7·87, say 8 inches), the ratio of the speed will be nearly 55 per cent. As to the quantity of action transmitted to the wheel, when it travels at this speed, it may also be admitted that the mean will be 60 per cent. of that which is due to the total fall, but that it will be somewhat less for the higher falls and smaller openings of sluices, and somewhat more for lower falls and larger openings."

This corresponds with Mr. Rennie's experiments, noticed in a former part of this book. The reader who wishes to investigate this subject, would do well to study General Poncelet's work, containing the "Second Mémoire" on experiments made on the large scale, as also the researches of MM. Poncelet and Lesbros on the flow of water.

In a discussion which took place at the Institution of Civil Engineers, when M. De Bergue produced the drawing here engraved, the only point noticed as requiring improvement was the masonry below the wheel; it was thought that the platform of the race should stop somewhat short of the vertical line, so that the rising buckets might be emptied as soon as they passed it, and thus avoid the possibility of retardation from their retaining any water after it had ceased to be useful.*

* The Poncelet water-wheel is not properly placed either in the class of undershot wheels, though the water be admitted "par dessous," nor in that of breast wheels. It is, in fact, a wheel of re-action, and the main object in view has been thus to fulfil in a wheel under the structural circumstances of an undershot the grand condition of all good machinery recipient of water power, viz., that the water shall *reach the wheel without shock, and quit it without relative velocity.*

CHAPTER XI.

THE WATER-PRESSURE ENGINE AND THE WATER-RAM.

HAVING considered the means of obtaining mechanical power with a rotary motion, it is now proposed to describe another mode of employing the power of water in a manner essentially different, but not less useful and important, which has been too much neglected in this country. The advantages to be derived from the use of the WATER-PRESSURE ENGINE, in hilly districts, where high heads and an abundant supply of water are found, especially when mines exist in the same locality, and require to be drained in order to work out the ore, or when power is wanted to raise the ore from the mines, seems so obvious, that it is strange to think such ready and effectual help to the miner should be passed by; for the reciprocating motion of the pressure-engines may not only be applied directly to the pumps, but may also, by using cranks and fly-wheels like those of the steam-engine, be made to turn the winding machinery and the crushing mills, or any other revolving mechanism. Thus, by the pressure-engine and the turbine, the power of waterfalls of any height, however great, may at once be made available.

The first invention of the water-pressure engine, like many other mechanical contrivances, appears to belong to Germany, and probably had its origin in Hungary, where so many ingenious machines, actuated by water, have long been used.

In the pressure-engine the power is obtained from a descending column of water acting by its weight or hydrostatic pressure upon the piston of a cylinder, in the same way that steam is employed; but as steam is an elastic fluid, and water (practically speaking) is not, care must be taken to prevent any sudden check to the motion of the water, which might break or damage the engine, and provision must be made to relieve the mechanism from any accidental shock so produced. In other respects, the action and the application of the water nearly resemble those of high-pressure steam.

The Germans appear to have made successive improvements upon their original engines, and to have extended, from time to time, their usefulness and employment, of which two important examples may be given.

The one is at Illsang, in Bavaria, at the salt-works, which are situated in the southern part of the kingdom. These works are supplied from a mine of rock-salt in the valley of Bergtesgaden, and from the salt-springs in Reichenhall, where the salt was formerly purified by solution and evaporation; but as this operation could not be carried on with advantage, on account of the scarcity of fuel, the saturated brine is now conveyed by a line of pipes 7 inches in diameter, through which it is forced from stage to stage for a distance of about 60 miles by a series of nine pressure-engines, acted upon by falls of water from the hills, and each of them working a pump.

A description of these engines will be found in the Proceedings of the Institution of Civil Engineers, and an excellent drawing, by Mr. W. L. Baker, of one of the best engines of the series, constructed by M. de Reichenbach, is in the collection of the Institution. This engine has a cylinder of 26 inches in diameter, with a stroke of 4 feet, making, in regular work, five strokes per minute; it is made entirely of brass, and is an exceedingly good machine both in design and workmanship. Very few working parts are visible, and it acts almost without noise, the sliding valves, or rather sliding pistons, which regulate the engine's action, being also moved by water pressure.

The other example, at Freyberg, in Saxony, is an engine constructed, in the year 1824, by MM. Brendal, for draining the Alte Mördgrube mine. It has two single acting cylinders attached to opposite ends of a working beam, by means of arched heads and chains; the cylinders are open at top, and have strong piston-rods of timber. The pressure of the water acts alternately under the piston of either cylinder, and forces it upwards, whilst the piston of the cylinder at the other end of the bearer is depressed by the weight of the pump-rods. A bell crank, attached to each piston-rod, gives motion to the pump-rods, each working twenty-two pumps, placed one above the other, lying at an angle of forty-five degrees, and dividing the lift of each set of pumps into twenty-two heights or stages of about 30 feet. The engine is placed 360 feet under ground. The cylinders are of cast-iron, 18 inches in diameter, with a stroke of 9 feet, and the useful effect was computed by M. von Gerstner to be 70 per cent. of the power expended. A section of one of these cylinders has been sent to the author by a friend at Wiesbaden, and a very complete drawing of this engine by Mr. Baker will be found at the Institution of Civil Engineers.

The first water-pressure engine used in England was erected by Mr. William Westgarth, at a lead mine belonging to Sir Walter Blacket, in the county of Northumberland, in the year 1765.

The cylinder of this engine was equal in length to the whole height of the fall of water; it was open at the top, and the water ran into the open top of the cylinder by a trough; the piston worked in a bored chamber, at the lower end of the cylinder, of 10 inches in diameter, and was attached by a chain to the arched head of an engine-beam placed above, the opposite end of the beam suspending a wooden rod, which passed down the pit to work the pump.

The column of water always pressed upon the top of the piston, but by admitting water below the piston, the pressure was neutralised, and the piston was raised by the weight of the descending pump-rods.

On closing the communication with the underside of the piston, and discharging the water from the cylinder bottom, the pressure of the column again acted upon the piston and sent it down. By a simple self-acting piece of mechanism, similar to the working-gear of the steam-engines of that time, the orifices were alternately opened and shut, and the reciprocating motion of the engine continued.

A detailed account, with drawings of this engine, was submitted to the Society of Arts, in the year 1769, and printed in the fifth volume of the Society's Transactions in 1787. The description and drawings were made by Mr. Smeaton, and a working-model then constructed from them is in the Society's possession. The Society voted fifty guineas to Mr. Westgarth, and presented a silver medal, with their thanks, to Mr. Smeaton, for his excellent account of so valuable an invention.

The author has carefully examined these interesting memorials of the early encouragement given to inventive genius by the Society of Arts, which continues with unwearying energy its career of public usefulness. He is satisfied, not only from the evidence which the machine itself offers, differing as it does entirely from any of the German engines, but from the written testimony of Mr. Smeaton, that Mr. Westgarth's pressure-engine was his own invention; and that he borrowed no part of it, either in plan or in detail, from anything then or previously existing on the Continent. His idea appears to have been taken from the single-acting open-topped atmospheric engine of the period; substituting the pressure of a column of water for the pressure of the atmosphere.

The liberality and goodness of heart which distinguished Mr. Smeaton, no less than his high talent and skill as a civil engineer, appear conspicuously in the correspondence alluded to, as will be seen in the following short extracts:—

"*Austhorpe, April* 29, 1769.

"I had the pleasure of seeing the first complete engine of this kind at work in the summer of 1765, for draining or unwatering a lead mine belonging to Sir Walter Blacket, at Caldcleugh, in the county of Northumberland: since which, that machine has been shown to all those that had the curiosity to see it. Mr. Westgarth has now erected four others in the different mines of that neighbourhood, one of which I have seen, and all attended with equal success."

In a subsequent letter, Mr. Smeaton says:—

"Mr. Westgarth was induced to think of applying for a patent for the exclusive privilege of using this invention, but previous thereto, he was pleased to advise with me concerning it, being frequently at that time in those parts of the country as an agent of Greenwich Hospital.

"Much as I admired the ingenuity of Mr. Westgarth's invention, I dissuaded him from the thoughts of a patent, as it would take a length of time to be sufficiently known, and the number of cases in which it could be properly applied were not sufficient to afford such a number of premiums as might defray the expense of a patent, with a prospect of advantage to himself and family.

"I therefore recommended it to him, as the Society for the encouragement of Arts, Manufactures, and Commerce were, by handsome premiums and bounties, encouragers of all useful inventions and improvements, to communicate his invention to them, that it might be made public, in confidence that he would obtain a bounty for the same, of such value as to the Society should seem meet: and in consequence hereof, I gave him a representation of the utility of his invention, with which, in the year 1769, he applied to the said Society, and obtained a bounty upon condition he delivered to the Society a working-model and a draught showing the construction of the engine: but as the death of Mr. Westgarth, which happened not long after, prevented the usefulness of the machine from being so successfully spread as it doubtless otherwise would have been; and as some of the most essential parts of the machine cannot be seen in the model without taking it to pieces; and the drawing not being accompanied with any literal explanation, nor the details of it sufficiently made out; at the request of the Society, I have now supplied these defects, that it may be published in such a manner that the utility of it may be seen, and the means of making and applying it be explained."

It appears that on seeing Mr. Westgarth's engine, Mr. Smeaton suggested to him that if the engine, instead of the great lever or balance beam, were made to work with a wheel, and instead of the long spear going down the descending pump-trees, the pistons were made to communicate with the main chain through a collar of leathers or stuffed collar, that then the whole of the machinery would stand together, just above the level or sough, and there would take up less room, as the work might in general be comprehended within the limits of the shaft; and the descending pipe might also be of a less bore, and be fixed in the corner of the shaft, and, therefore, be upon the whole much more convenient for underground works; and on this mode the latest engines of Mr. Westgarth were actually constructed with success. So that little was wanting except the perfection of modern workmanship, to render the engine complete, although it was not made direct-acting, as the engines recently made generally are.

Mr. Smeaton constructed a water-pressure engine, in 1770, at Temple Newsam, Yorkshire, to work a pump for supplying Lord Irwin's residence. It was, of course, of small size; and the pressure was communicated to the cylinder by an inclined pipe, bringing the water from some distance. He modified and improved the plan of Mr. Westgarth, closing the cylindrical tube, and using the piston-rod with a stuffed collar; still, however, retaining in its original form, or nearly so, the ingenious slide-valve designed by Mr. Westgarth.

This was a cylindrical hoop or ring, sliding, for a short distance, up and down a pipe which it encircled, the pipe having two sets of openings, separated by a horizontal bridge or partition. The valve was enclosed in a box attached to the branch pipe of the cylinder; it sustained the pressure of the columns equally in all directions; and it was rendered water-tight by strips of leather.

When the upper openings in the pipe were exposed above

the water, the water entered the cylinder below the piston; but when the lower set of apertures was opened, by raising the cylindrical valve, the water escaped from the cylinder;

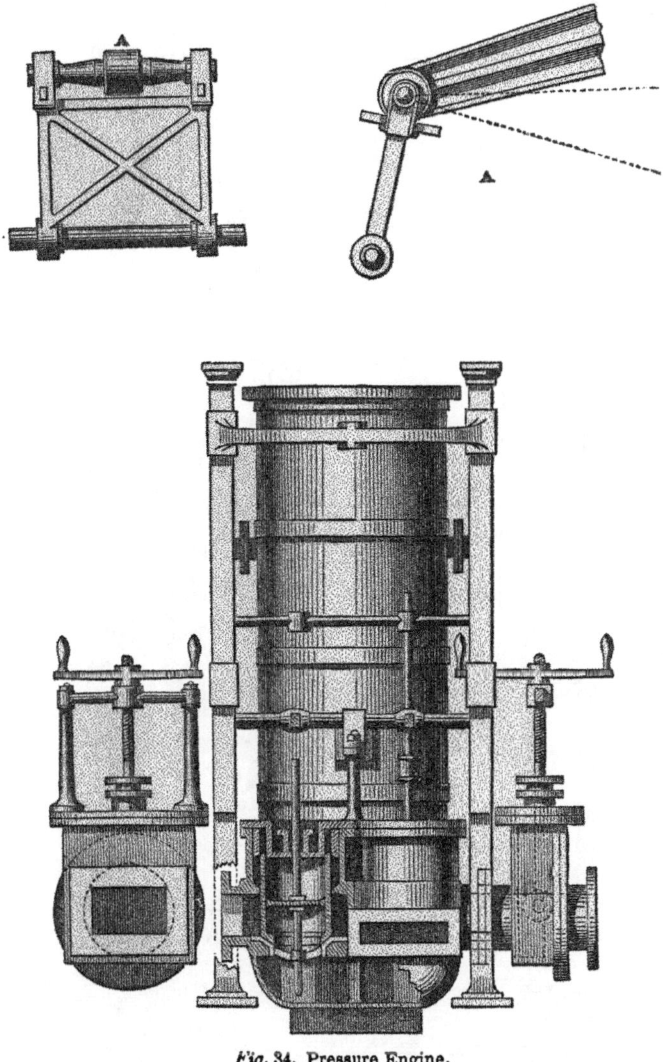

Fig. 34. Pressure Engine.

the position of the valve at the same time shutting off the further entrance of the water and its pressure upon the piston. The openings were, in the first instance, square

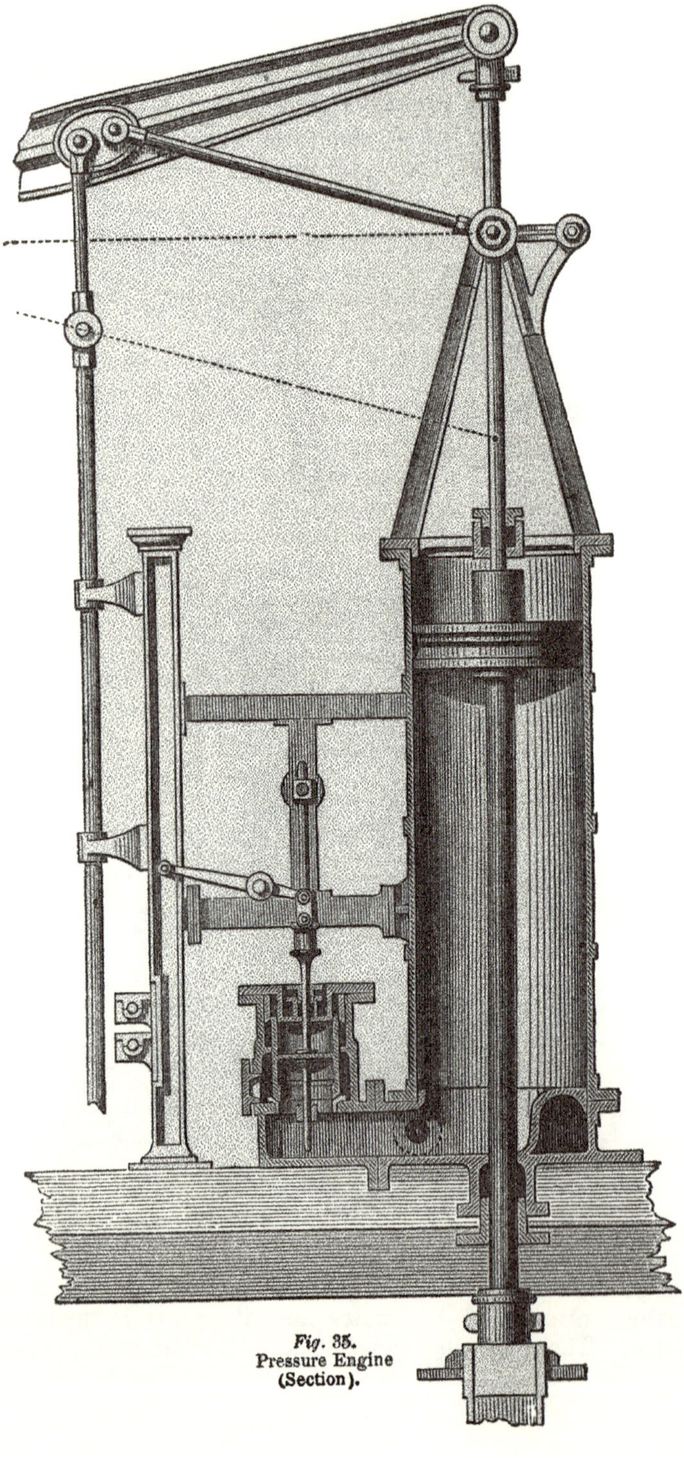

Fig. 35.
Pressure Engine
(Section).

holes; but afterwards they were made in the form of a lozenge or rhombus, with the acute angle upwards, so that the water might enter and be shut off more gradually.

After Mr. Smeaton's time, the water-pressure engine seems to have remained in abeyance, and none were made until Mr. Trevitheck revived their use. The great improvements made in the steam-engine by Mr. Watt, caused water engines of all kinds to be neglected, and even water-wheels were, in many cases, superseded by steam-engines. Water-power went out of fashion, and was generally considered to be too precarious and expensive, as compared with steam, to deserve much attention from engineers.

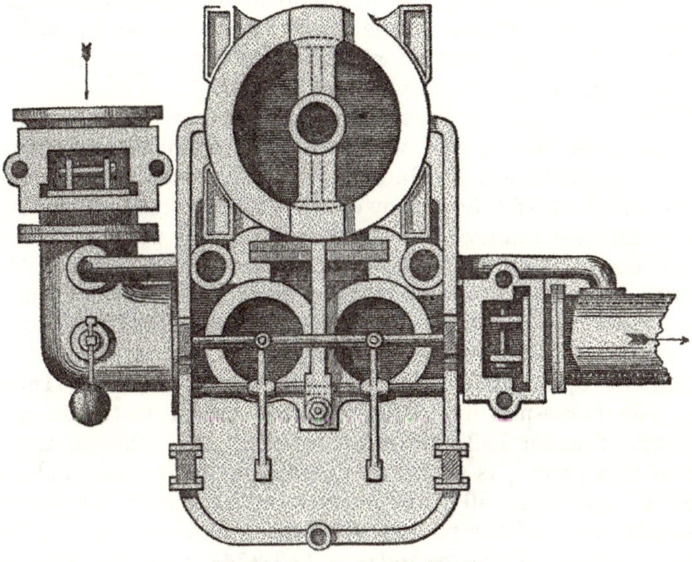

Fig. 36. Pressure Engine (Plan).

Lately, however, more enlarged views have been taken. Men who stand high in their profession have turned their attention to this subject, and it has been more studied and better understood.

Mr. Trevitheck constructed several water-pressure engines, one of which was erected in Derbyshire, in 1803, and is still at work at the Alport mines. The cylinder of this engine is 30 inches in diameter.

In 1841, Mr. John Taylor, well known for his long experience in mining operations, and whose application of water-wheels to the working of pumps has already been mentioned, advised the employment of another and more powerful engine at the Alport mines, near Bakewell, which was made under the author's direction, at the Butterley works. This was the most powerful engine of the kind that had been made. The cylinder is 50 inches in diameter, and the stroke 10 feet. It is worked by a column of water 132 feet high, acting below the piston, and lifting, by direct action, a weighted plunger-pole, 42 inches in diameter, which raises the water from the mine to a height of 132 feet; so that the proportion of power to effect is as the area of the piston to that of the plunger: namely, 1,963 to 1,385, or full 70 per cent. (See Figs. 34, 35, and 36.)

Mr. Darlington, the superintendent of the machinery at the mines, fixed this engine in the shaft; in a letter to the author he states that the cost of maintaining the engine has been less than £12 a year since it was erected.

The usual speed is about five strokes per minute, but it will work at the rate of seven strokes, without any concussion in the descending column; the duty actually done being then equal to 168 horse-power of 33,000 pounds raised one foot high in a minute. Thus the area of the plunger, 3 feet 6 inches in diameter, is 9·621 square feet × 10 feet, the length of stroke, × seven strokes per minute = 673·47 cubic feet raised 132 feet high in a minute; and 673·47 cubic feet of water × 62·5 pounds, the weight per foot × 132 feet in height = 5,556,127, which, divided by 33,000, gives 168⅓, say 168 horse-power. The pressure upon a piston from a column of water 132 feet high, reckoning 27 inches of water equal to a pound, is about 58 pounds on the square inch, or rather more than 50 tons' pressure on the area of the piston. Thus the area of the piston, 50 inches in diameter, is 1,963 square inches × 58 pounds = 113,854: and this, divided by 2,240, the number of pounds in a ton, gives 50·8, say 50 tons.

There are five other pressure-engines at these mines of smaller size.

This engine was erected early in 1842, and is now working. It was at work, without intermission, for six years; and on one occasion, when the author made inquiry as to its performance, the answer was, that it had been constantly going for the last seventeen weeks, and nobody had seen it during the time.

An excellent working model of this engine will be found in the Museum of Practical Geology in Jermyn-street, London, an institution which ought to be visited by all persons who are interested in mining and metallurgy, and in the employment of mineral wealth.

It will be observed that after the large valves are closed, the pressure is continued upon the piston to complete the stroke. This was at first done by means of cocks; but as the friction of these caused some little trouble, Mr. Darlington substituted some small pistons to shut off the water at the termination of the stroke.

Mr. Taylor has since had another engine of the same size made for a lead mine in Wales, and the results have been equally satisfactory.*

In these machines, as in all others where the water acts by its gravity or pressure, the best results are obtained where the water enters them without shock or impulse, and quits them without velocity. We then realise all the available power the water will yield with the least loss of effect; and the duty is best performed by making the pipes and passages of sufficient and ample size to prevent acceleration of the hydrostatic column.

The cranes worked by water-pressure, constructed by Sir William George Armstrong, of Newcastle-on-Tyne, have been described in another book, on *Cranes and Hoisting Machinery*, written by the author of this work.

These hydraulic cranes are characterised by their great steadiness and precision of movement. By means of the regulating handles, their motions, both in lifting and lowering, as well as in turning, are graduated with perfect accuracy; and, practically speaking, the speed with which they may be worked has no other limit than that imposed by the size of the supply-pipe.

The author has the satisfaction to state that since the treatise on Cranes was published, many of those worked by water-pressure have been erected in and about London,—particularly in the collier basin of the West India Docks, where they are used to discharge the cargoes of ships, and at the Great Western Railway goods station. They are also used at several other railway goods stations.

Hydraulic pressure admits of very extensive and useful application in mercantile docks, not only to cranes for lifting

* One of the most remarkable and complete water-pressure engines is the large one employed at Huelgoat, in Brittany, of which an account may be found in the "Annales des Mines," and also in "Les Machines destinés a l'elevation des eaux" of General Morin.

heavy weights, but also for "*whipping*" light goods from ships, and for opening and shutting dock-gates, swinging-bridges, and sluices.

The facility with which water may be conveyed to the places where the power is wanted, the ease with which it may be managed, its perfect safety, and its constant readiness for action, render it eminently suitable for these purposes.

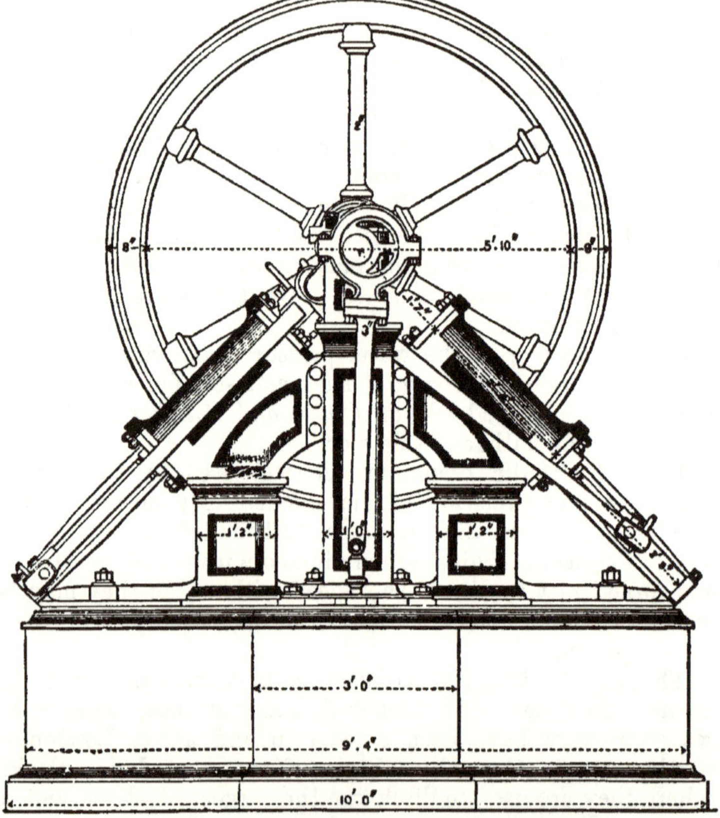

Fig. 37. Armstrong's Water-Pressure Engine (Elevation).

When hydraulic engines are extensively used and systematically employed, it may be prudent to apply steam power to raise the water to an elevated tank, not less than 100 feet above their level, for that is in reality only another mode of using steam power, the water being the vehicle for its transmission. But there are many places where the supply may with advantage be obtained from the public water-works, and even cases when the water pressure may be employed *in transitu*, without using the water. This is now actually done in Newcastle-on-Tyne, where an improved pressure-engine, made by Sir W. Armstrong, is used to print the *Newcastle Chronicle*, one of the most extensively circulated provincial newspapers. The author had the pleasure of seeing it in full work. The town is placed on the side of a high and steep hill, and the printing-office is about half-way up, so that the water merely passes through the engine and goes on to aid the supply for the lower parts of the town.

In many parts of London water is supplied, in quantity, at 4*d.* for 1,000 gallons, at a pressure of 150 feet; and there are many trades that require occasionally to use mechanical power, but cannot employ a steam-engine, for they do not want the power constantly, and they do not want much power at any time.

If the supply-pipe, conveying the water from the main into the workshop be of sufficient size, a small pressure-engine may at any time be set in motion, without risk of fire, and without requiring any skill in management, by the mere turning of a tap; and the turner, cutler, optician, or other artificer, may have auxiliary power whenever it suits him to use it—the engine being also a perfect water-meter, when a counter is attached to it, shows the number of revolutions it has made, and the quantity of water that has passed through it. A gallon of water weighs 10 lbs., so that 1,000 gallons of water falling 150 feet are equal to 1,500,000 lbs. falling one foot; and, if 1,500 gallons of water be used in one hour, they are equal to 37,500 lbs. falling one foot in a minute, or somewhat more than one horse-power, which is 33,000; therefore, allowing the difference for loss at the escape-valves, it may be assumed that the cost of one horse-power for one hour will be 6*d.* where the water runs to waste.

The engine which prints the *Newcastle Chronicle* has two cylinders, each $3\frac{3}{4}$ inches in diameter, or 8·9461 square inches area on their pistons; but the piston-rods are each one inch in diameter, and their area, ·7854 × 2, must be deducted, leaving the area of the four sides of the pistons, less that of the rods, 34·2136 inches. The length of stroke is 7 inches, and the engine makes from 55 to 60 revolutions per minute, so that taking the gallon of water to be 277·25 cubic inches, the total quantity of water used in one hour to work this engine is from 2,850 to 3,110, say 3,000, gallons per hour. But in this case it is the pressure alone that is used, and not the water. The whole head of water is 430 feet, and of this column 200 feet are above the engine, and 230 feet below it. This difference of 200 feet, allowing 27 inches of head to give 1 lb. pressure, is equal to 88 lbs. on each square inch of the pistons. The stroke being 7 inches, a speed of 60 revolutions is equal to 420 inches, or 35 feet, per minute; so that, multiplying the area 34·2 inches × 88 lbs. × 35 feet per minute, the product is 105,336 lbs. falling one foot in a minute, which, being divided by 33,000, gives 3·19, or full three horse-power.

Sir W. Armstrong has made a water-pressure engine for an underground inclined plane in South Hetton Colliery. It has four cylinders, each only 3 inches in diameter, and 12 inches stroke. It usually works at about 100 revolutions per minute, and it is quite free from concussion at this high speed. The column of water acting upon it is 600 feet, and the diameter of the supply-pipe is 4 inches. The engine is placed about 300 yards from the bottom of the shaft at the head of the inclined plane, 880 yards long, of an irregular acclivity, but in the steepest parts the rise is 1 in 18. Up this incline the engine hauls twenty loaded waggons, weighing collectively 15 tons. Each run is accomplished in six minutes, and the quantity of water expended in that time is 1,500 gallons. This engine will eventually be required to draw forty trains per day, in which case the quantity of water expended will be equivalent to a constant feeder of 42 gallons per minute. The greater part of the water used by this engine drains out from the upper part of the shaft, and before the engine was erected it was permitted to fall to the bottom, but it is now collected into a reservoir for the supply of the engine; and when it is required to draw forty trains per day, the increase of pumping to be done by the great lifting engine which drains the mine will not exceed 20 gallons per minute during the twenty-four hours. The work done by this engine appears to be from 30 to 35 horse-power.

Another engine, with four cylinders, is now in use for winding up ore, &c., from a lead mine at Allenheads. The cylinders are 6 inches in diameter, with a stroke of 18 inches; the column of water acting upon it is about 230 feet in height, and the expended water escapes from the mine by a level: it commonly makes about sixty revolutions per minute, and then it raises a load varying from 7 to 10 cwt., exclusive of

Fig. 38.

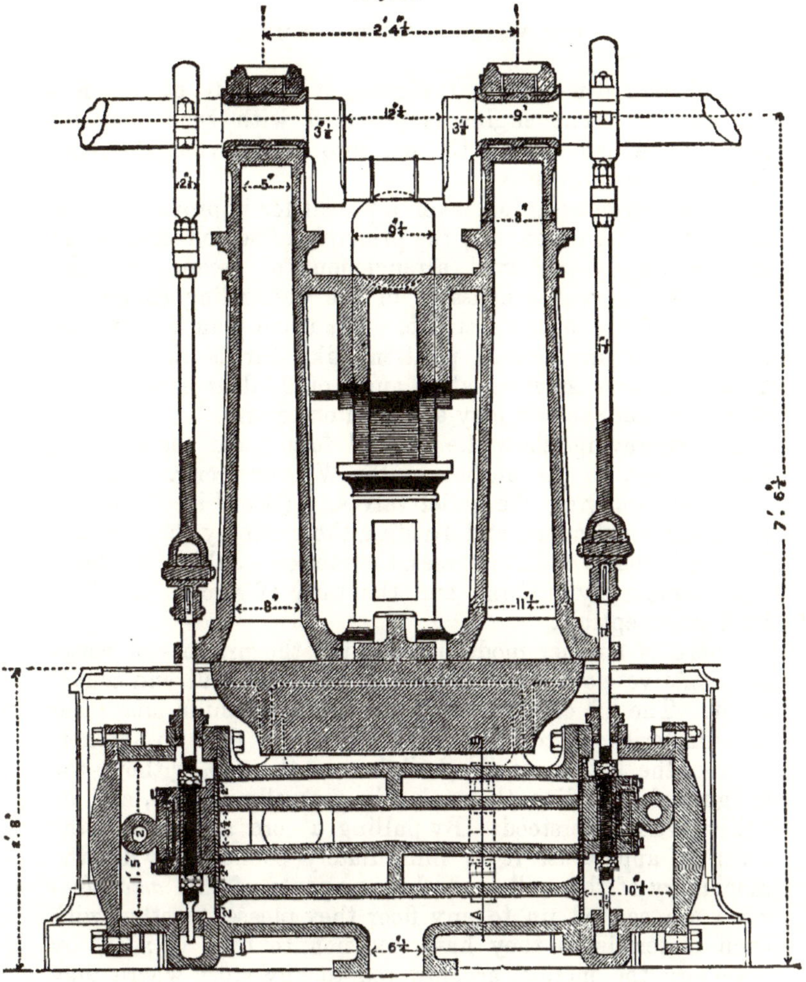

Transverse Section.

Side View of Slide Valves.

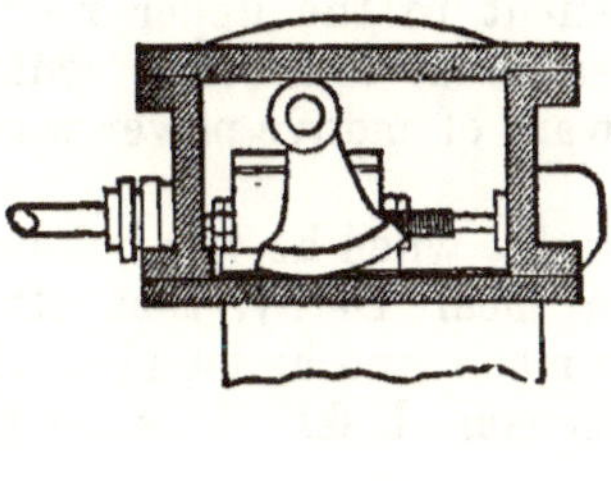

Section through A B.

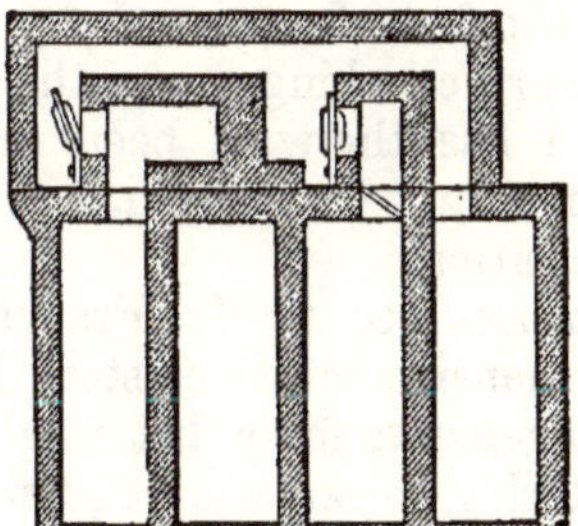

Relief Valves.

the unbalanced weight of the rope, at the rate of 900 feet per minute. The engine is provided with spur gear for pumping, which can be thrown into action when the rope-drums are not in use.

Sir W. Armstrong has taken a patent for improvements on the water-pressure engine, the principle of which consists in an ingeniously-contrived arrangement of escape or relief valves, to prevent concussion of the descending column at the turn of the piston's stroke. His usual plan is to fix the cylinders in an inclined position, like Sir Mark Brunel's *Thames Tunnel Engines*, at an angle of 90 degrees with each other, so that the two may act upon one crank.

The engravings show the general form; and the sections, kindly given to the author by Sir W. Armstrong, show the mode of applying the relief-valves. (See Fig. 38.) By using three cylinders, and dividing the circle by three cranks into equal parts of 120 degrees, the motion would be rendered almost perfectly uniform, and the turn of stroke would be hardly perceptible.

There is another mode of applying the pressure of water as a motive power, which is exceedingly useful and convenient. The lift or hoist employed in cotton-mills and other manufactories to convey persons from one floor to another, or from the bottom of the building to the top, without the fatigue of ascending the stairs, is generally known, and its utility well understood. By pulling a cord attached to the hoisting apparatus, it is immediately connected with the machinery of the mill, and the occupants of this *ascending-room* are carried up to any floor they please, together with such materials as they have occasion to take up; or, by reversing the motion, are lowered to any story where they wish to land.

Such a lift would be a valuable addition to the convenience and comfort of many a lofty house in London, and save many a weary climbing from the basement to the upper rooms. Often has the wish been expressed, that some such contrivance might be adopted, but the want of motive power was a prohibition.

It occurred to Mr. Francis Wright, when he was building his mansion of Osmaston Manor, near Derby, that other arrangements for a lift might be made, and he found at the Butterley works the requisite mechanical skill to carry his ideas into practice.

At a convenient height above the manor was a reservoir of

some extent, which was constantly replenished by means of a water-wheel in the park working a set of pumps, so that a supply of water at considerable pressure was always available.

A cylinder of cast-iron, about 46 feet long, was made in convenient lengths, and truly bored to an internal diameter of 11 inches: into this a piston was fitted and made water-tight by a leather collar, such as is used in the hydraulic press. The cylinder was sunk in the ground, and set accurately upright in the basement of the house, and the stem of the piston carried a platform or landing, with suitable railing and other appliances.

A pipe leading down from the reservoir conveys the water into the cylinder, below the piston, which forces it up with the platform and the persons upon it, until the ascent is stopped by shutting off the water; then, by allowing the water to escape from the cylinder, the piston and platform descend by their own weight until the escape of the water is prevented, or until the piston reaches the bottom of the cylinder.

The ascent required at Osmaston Manor is $43\frac{1}{2}$ feet; four persons are carried up in two minutes to the full height; and as the platform cannot rise or fall faster than the water enters or escapes, there is no risk from acceleration either way, because the size of the water-way determines the maximum speed of the piston, which is also regulated by the opening given to the valve.

The apparatus has been in action for several years, and has answered its purpose well. A similar machine, with a shorter lift, is used to raise the provisions from the kitchen to the level of the dining-room.

Assuming the elevation of the pond or reservoir to be 130 feet above the cylinder bottom (which is about the height), the pressure against the piston at the commencement of the lift will be about 58 lbs. on the square inch, and when it has risen to the full height of $43\frac{1}{2}$ feet, the pressure will be about 43 lbs. on the square inch, at the conclusion of the lift. The area of a piston 11 inches in diameter, being about 95 square inches, a pressure of 58 lbs. gives a force of 48 cwt., and a pressure of 43 lbs. gives a force of 36 cwt., which is found sufficient to overcome the friction, resistance, and weight of the machinery, and to raise the requisite loads, carrying up half-a-ton in about four minutes.

THE WATER-RAM.*

In treating of water-pressure engines, it is remarked that care must be taken to prevent any sudden check being given to the descending column, and that relief or escape valves, air-vessels, weighted plungers, or some means of lessening the momentum of the water, are requisite to guard against the shocks which the machine must sustain if the motion of the water be impeded: this momentum, however, may be rendered exceedingly useful as a moving power, by which a portion of the water may be raised to a higher level by means of *the hydraulic ram*, to be used for manufacturing and domestic purposes, or for watering land above the level of the stream.

The water-ram, although it often figures in books on hydraulics, is less used than it ought to be: the reader generally regards it as a curious, old-fashioned device, and thinks no more of it; but for simplicity, cheapness, and useful effect combined, few machines are superior.

The merit of first employing the momentum of water to raise a portion of itself to a higher level is due to Mr. John Whitehurst, who, in a letter addressed to the Royal Society, in 1775, and printed in their Transactions, explains the principles of its action, and the mode of applying them, he had then successfully adopted.

M. Montgolfier, who does not appear to have known anything of this communication, afterwards invented a similar machine in France, and gave it the name it still bears, "Le Bélier Hydraulique," from the butting action of the water; and Messrs. Boulton and Watt, with his permission, took a patent for it in England, dated December 13, 1797.

About twenty years later, the son of M. Montgolfier took an English patent for some improvements in his father's arrangements; and the visitors of the Great Exhibition of

* Everything that has been so skilfully and successfully carried into use upon an extensive scale by Sir William Armstrong in the transference and modification of power by the intervention of liquid connectors, is originally due to the inventive ability of Joseph Bramah, the inventor of the hydraulic press itself. No detail or possible form of application seems to have escaped him. Documents in his own hand exist which prove that he had anticipated and even tried to get introduced all the varied methods for working docks, and dock traffic, &c., as since carried out by Sir William Armstrong.

1851 will remember two modifications of the hydraulic ram by London makers—Mr. Freeman Roe and Messrs. Easton and Amos.

The first machines made on the plan of the elder M. Montgolfier, by Messrs. Boulton and Watt, were exceedingly simple in their construction and arrangement, and yet very effective. A horizontal pipe, with a trumpet-shaped mouth to admit the water freely, was either inserted into the side of a supply-tank into which the stream flowed, and kept it full, or it passed through the dam into a reservoir that could be maintained at a uniform height. At the end of this pipe was a weighted valve, opening inwards, and upon the top of the pipe, near the end, was an air-vessel, the capacity of it being at least equal to ten times the quantity of water to be raised at each stroke, and larger as the lift increased. At the bottom of the air-vessel was a valve, opening upwards, to admit the water into the air-vessel, and prevent its return, and to the air-vessel was attached the ascending-pipe, carrying up the water, raised to a higher level by the action of the machine.

The water being allowed to flow through the horizontal pipe, soon acquired velocity and force, and suddenly shut the valve at the end of the pipe; the momentum of the water, thus checked, lifted the valve in the air-vessel, forcing through it a portion of the water, compressing the air within, and thus raising the water part way up the ascending pipe.

The force having been expended and the water brought to a state of rest, the weight-valve at the pipe end opened again; the descending column flowed through until its velocity again overcame the weight on the valve and forcibly shut it: thus, by repeated strokes, the water was raised to the top of the ascending pipe, and then every stroke discharged a volume of water inversely proportioned to the height.

The pipe conveying the descending water may be placed in an inclined position, instead of being laid horizontally; it is sometimes called the body of the ram, and the air-vessel and valves the head. M. Montgolfier considered, that with a well-constructed machine he might obtain 75 per cent. of useful effect; but the following were the results of experiment:—

The fall of water was 3 feet 4 inches, the pipe, or body of the ram, 8 inches in diameter, and the height to which the water was raised, 15 feet 1 inch. The machine made 100 strokes in three minutes, expending in that time 67 cubic feet of water, and lifting 9¼ feet. Thus 67 × 3·33 = 223·11, and 9·25 × 15·083 = 139·517. Then $\frac{139 \cdot 517}{223 \cdot 11} = \frac{62}{100}$, or 62 per cent. of the power employed. The following table shows the results obtained by French engineers.

The first column shows the head or fall of water acting on the machine; the second, the quantity expended. The third shows the height to which the water was raised; and the fourth the quantity raised to that height. The fifth column shows the useful effect in each case.

Fall in feet.	Expended.	Elevation.	Raised.	Effects.
8·53	2·405	52·69 ft.	·2207	·565
27·30	5·951	195·02	·6189	·543
34·77	2·971	111·88	·6013	·651
3·21	70·282	14·92	9·5147	·629
22·96	·459	196·86	·0343	·640

The quantities of water expended and raised are reduced to cubic feet, and the mean of useful effect is ·605, or say six-tenths of the power expended: experiments also show that it can do good duty where it raises water seven times the height of the fall. Practically, perhaps one-half of the power expended, or 50 per cent. of useful effect will be the amount realised; for it must be expected that in all these experiments the machines were in good order; yet there are many localities where a water-ram may be used with great benefit, as it steadily and constantly performs its task with very little care on the part of its owner, except to see that the water is strained through a grating, to keep back extraneous substances, and that the apparatus is protected or emptied during a hard frost. The first example mentioned raised 9¼ cubic feet, or 57·8 gallons in three minutes, amounting to 27,750 gallons in twenty-four hours, to a height of 11 feet 9 inches above the pond that supplied the water; and this without labour or expense, beyond the interest of its cost.

It is true that the heavy blow or shock, given by the momentum of the descending column, has hitherto limited the size of this apparatus; but this shock may be lessened by increasing the number and improving the form of valves and pipes; and the principle may probably be still further developed in the hands of skilful mechanics, as it has been successfully done in the pressure-engine.

After the foregoing passages had been written, the author was favoured with a letter and drawing from Messrs. Easton

& Amos, of Great Guildford Street, Southwark, who exhibited one of their machines at the Crystal Palace; the drawing showing their mode of fixing it, to supply water to a mansion in the country.

They have made some improvements, for which they have taken a patent; and they state they can obtain a mechanical effect of 65 or 70 per cent.; that the lowest available fall is 2 feet, and the highest 35 to 40 feet; the quantity of water they lift varies with the size of the ram and height of the lift from half a gallon to 6 gallons per minute.

The greatest height to which they have raised the water, by a machine which has been at work some years, is 330 feet. They lay the "*injection pipe*," or body of the ram, at an inclination varying from one in eighteen for small falls to one in four for high falls. Having formed a dam across the stream, they take the water by an earthenware conduit-pipe into a covered tank of brick-work; this receives the mouth of the inclined cast-iron pipe, which is opened or closed at pleasure by a flap valve; the pipe is laid in the ground below the reach of the frost, and the apparatus is fixed in a circular and domed vault of brick-work, which protects it from freezing; and from which the water that works the ram is conveyed by a covered channel to some convenient point where it may rejoin the stream at a lower level. The ram is secured to a stout piece of oak or elm timber built into the brick-work. (Fig. 39.)

In the preceding pages the reader will have observed that several machines derive their power from the reaction of water-pressure: such as Dr. Barker's mill, Whitelaw's mill, the vortex-wheel, and others. If the machines be impelled by some other power, and caused to revolve by an equal external force, say that of a steam-engine, they may be made to act as pumps; and as they had before been put in motion by the pressure of a column of water descending and passing through them, they would, by inverse action, raise a corresponding column of water to the like height.

Let the most simple of these machines, Dr. Barker's mill (fig. 16), be turned upside-down,—let the funnel mouth at the top, there shown as receiving the water, be immersed in a well, and the machine caused to revolve rapidly on its axis; the swift rotary motion will cause a partial vacuum in the arms, and the water will rise in the central pipe and fill them until it is thrown out at the holes near the ends, where the centrifugal force will cause continuous streams to be discharged so long as the requisite velocity is maintained.

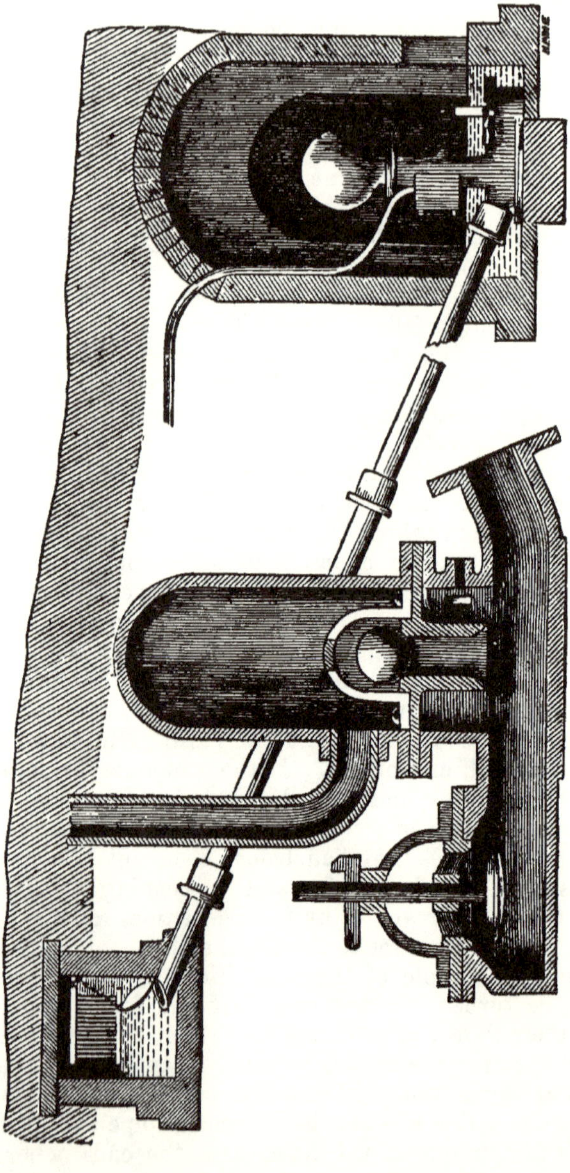

Fig. 39. Water ram by Easton and Amos.

The straight form of the arms, however, causes a considerable loss of effect: the course the water should take is that of the curve compounded of its radial direction, and of the rotary motion of the machine, for any radial velocity in the water, at the point of discharge, is power uselessly expended. Another centrifugal machine, having the same diameter, section, and apertures, but having the arms bent to the proper curvature, will discharge more than double the quantity of water in the same time with the same power—(see Figs. 18 and 20). This was proved by direct experiments made by Mr. Hensman, at the request of the jury, during the Exhibition of 1851.

Thus Mr. Whitelaw's mill will be found to make a very effective machine for raising water, by reversing its action; and hence it was the jury found that Mr. Appold's wheel, formed with vanes similarly curved, produced so much greater results than wheels of the same dimensions with straight vanes. Mr. Appold's wheel was only 12 inches in diameter; it received the water on each side, through apertures of 6 inches diameter, and had a central disc or diaphragm perpendicular to the axis, intersecting the vanes, forming, as it were, a double wheel revolving between two cheeks that projected from opposite sides of the reservoir.

The vortex wheel, perhaps, may serve to explain the form and operation of Mr. Appold's pump, by supposing the axis to be horizontal, and the water to enter at the sides and be discharged through the large round pipe. When this wheel was tried against two others of the same size, the one with straight arms or vanes, inclined at an angle of 45 degrees, and the other with radial arms, the following results were obtained:—

	Revolutions per minute.	Gallons raised per minute.	Height raised.	Useful effect.
Mr. Appold's wheel.	792	1164	18 ft. 8 in.	·649
" " .	788	1236	19 " 4 "	·680
Inclined vanes . .	694	560	18 " 0 "	·394
" " . .	690	736	18 " 0 "	·434
Radial vanes . . .	624	369	18 " 0 "	·232
" " . . .	720	474	18 " 0 "	·243

The experiments were made under the direction of General Morin, and the amount of motive power employed was ascertained by the dynamometer, constructed by M. Morin, on a principle proposed by General Poncelet.

The author was present during some of these trials, and was gratified to witness the care and skill with which they were conducted.

By admitting the water at both sides, the atmospheric pressure is neutralised and balanced; this is not the case in Whitelaw's arrangement, although a similar method has been used in other machines, as in the "Fan-Blast," for blowing furnaces and forges, several of which of large size have been constructed by the author, and machines, similar to the rotary fan, have also been applied to raise water.

At the close of the Exhibition of 1862 some rough experiments were made for the purpose of comparing the relative powers of the Gwynne and Appold centrifugal pumps which had been exhibited. In the interval which had elapsed since 1851 both pumps had been modified. In that year the pump made on Appold's system by Messrs. Easton, Amos, and Son, was considered to be superior to that of Gwynne's; but in 1862 the two pumps were much nearer to equality. The following results were obtained in 1862. The engines used for driving Gwynne's pump made 200 revolutions, which is equal to 466$\frac{2}{3}$ feet of piston movement per minute. The average pressure was 26·66 lbs. per square inch, and the indicated horse-power 190. Of this pressure about 21$\frac{1}{2}$ lbs. were available for driving the pump; this represents a horse-power of 154·12. The water lifted on an average 20·73 feet. The discharge was estimated at 91·03 tons per minute, which represents 128·1 horse-power as the amount of work done. In other words, 83·18 per cent. of the work developed was usefully employed.

In Easton, Amos, and Son's pump the piston moved through 208 feet per minute; the mean pressure was 29·522 lbs., and the indicated horse-power was 116·21. Of this 94·44 was estimated to be applicable to driving the pump, which made 124 revolutions per minute. The water was lifted 7 feet 1 inch, and the quantity discharged was estimated at 140·77 tons, which represents a horse-power of 67·68 given off in useful work; this yields a ratio of efficiency of 71$\frac{2}{3}$ per cent.

During a long practice in the drainage of extensive tracts of fen and marsh lands by steam-power, where natural drainage was impracticable, the author employed *scoop-wheels* to throw off the water. These were like the breast-wheel reversed, and the reader may imagine their action by referring to

Fig. 29, and supposing the wheel to lift the water, instead of being turned by it. Two wheels like this, 28 feet in diameter, driven by steam-engines, were constructed by the author, at the Butterley Iron-works, some years ago, for the drainage of Deeping Fen, near Spalding, containing about 25,000 acres, then often covered with water, but now growing corn. One of these is found sufficient, except in very rainy seasons, to keep the fen clear of water. It is turned by an engine of 80 horse-power, and the floats, or *ladle-boards*, travel at a mean rate of 6 feet in a second: these measure $5\frac{1}{2} \times 5$ feet, and deliver a constant stream of water, with a sectional area of $27\frac{1}{2}$ square feet, which moving at the speed above-named, discharges 165 cubic feet, equal to more than $4\frac{1}{2}$ tons of water in one second; or about 16,200 tons in an hour. A more simple or effectual mode of raising a large body of water to a height of 10 or 12 feet (from surface to surface) cannot well be devised, nor one less liable to derangement, from ice, weeds, and drift-wood. By this means upwards of 125,000 acres of fen land in England have been cleared of water under the author's direction, besides similar works of drainage in Holland, Germany, and in British Guiana, where the same machinery also irrigates the land in the dry season. For the drainage of small districts, it is probable that rotary or centrifugal pumps may be used with advantage.

CHAPTER XII.

THE APPLICATION OF WATER-POWER TO TUNNELLING.

We have long been accustomed to regard lofty and almost impassable ranges of mountains as effectual barriers for preventing any intimate intermingling between the peoples living on either side. The demands of commerce, and the advantages accruing from facile intercourse between distant places, have had a share in promoting the construction of railways across the plains and minor hills of Great Britain and the Continent, and probably it will not be long before the nations which are now separated by the Alps, &c., will be brought within easy reach of each other. At any rate, great efforts are being made to force a road through, or over, the rocky masses of the Alps, Pyrenees, and Apennines. These efforts have led to the construction of the great tunnel under Mont Cenis, which will be about 8 miles long: others, equally remarkable for their length, have been proposed. In order to carry out such gigantic works as these, it became necessary to contrive an improved method of tunnelling. The barren nature of the districts where these works would have to be carried on, and their difficult access, rendered an economical, yet rapid, method desirable. There is, however, another difficulty. In the tunnels hitherto constructed efficient ventilation could be secured by means of shafts, but as shafts are impracticable in driving a tunnel through the Alps, other means must be resorted to.

In the following chapter we propose to describe the machinery used for boring the Mont Cenis tunnel. The power employed is derived from torrents. Its application involves three operations, viz., (1) the compression of air by a pressure of 5 atmospheres, (2) the transmission of the air at this pressure to the end of the gallery, and (3) the economical utilisation of this pressure for working the machines. The air is compressed by two kinds of apparatus. In one the compression is effected by the direct fall of water, and in the other by means of a pump. The first kind is called the *compresseur à choc*, and is shown in Fig. 41. But we will first endeavour to explain the principles on which it acts. Fig. 40 is a theoretical diagram, and represents a tube shaped like an inverted syphon. The longer limb communicates with a reservoir of water, R, placed 26 mètres above the horizontal branch. The shorter limb communicates with a chamber filled with air, having an effective pressure of 5 atmospheres. At A is a valve for closing the longer limb, so as to intercept the communication with the reservoir R. At B is a valve opening upwards. At C is another valve for establishing or intercepting the communication with the compressed air in the chamber R′. At D is a fourth valve, which so opens as to allow fresh air to enter the smaller limb should a vacuum be formed in it. The air in the reservoir

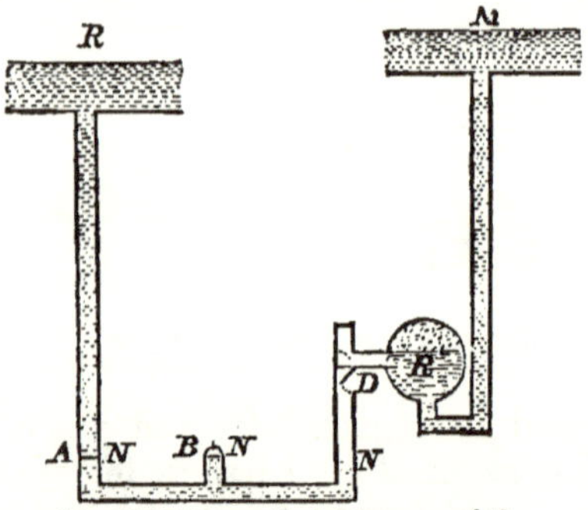

Fig. 40. Theoretical Diagram of the *Compresseur à Choc.*

R is kept at a uniform pressure of 5 atmospheres by a manometer fed by a small reservoir M, the water in which is maintained at a constant level of 50 mètres above the apparatus. If the valve A be closed, and the valve B opened, the pressure

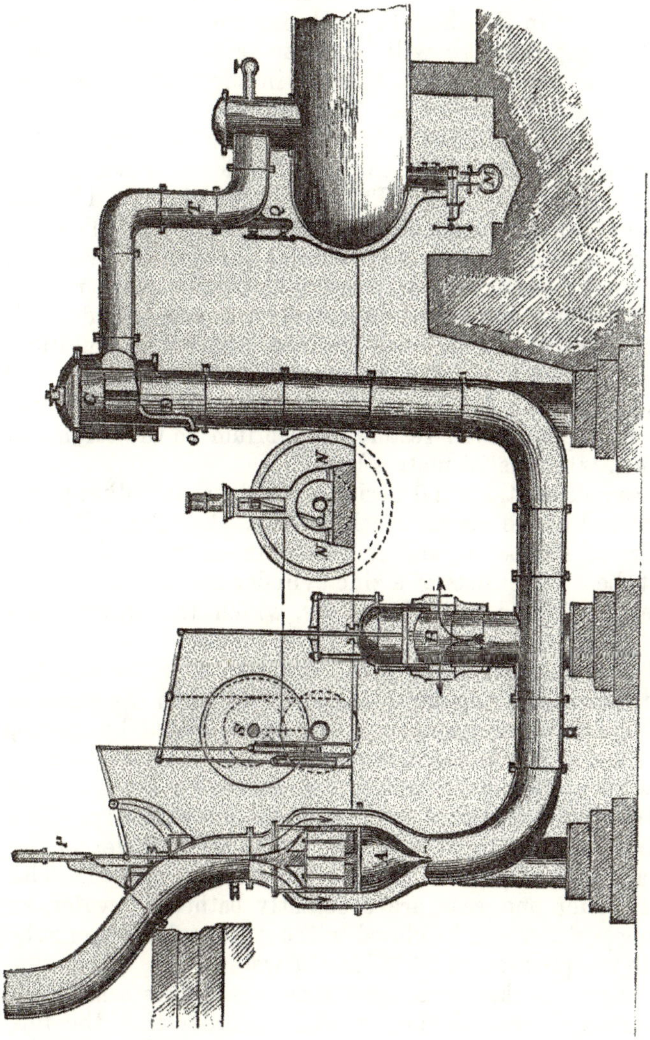

Fig. 41. Compresseur à Choc.

of the air in the chamber R′ will keep the valve C closed. The valve D will open and allow atmospheric air to flow into the small limb of the syphon up to the level N N, while water will flow out through the valve B. The lower part of the apparatus will consequently be filled with water up to the level N N. If now the valve B be closed and A opened, the column of water R A will descend, and press upon the water remaining in the horizontal part of the syphon, and this will, in its turn, press against the air in the short limb, closing the valve D. Whereupon the compressed air will open the valve C, and penetrate into the reservoir R′, so that the air in the reservoir will be under a greater pressure than 5 atmospheres.

If the valve A be closed when all the air has entered the reservoir the descent of the column of water will be arrested, and the valve B will open. The valve C will be closed by the pressure of the air inside the reservoir R′, water will flow out at B, and so allow fresh air to enter through the valve D. The power of the machine therefore depends on the rate at which the water descends in the tube R A, and this depends upon the height. The rate obtained under the conditions above mentioned will suffice to impart an effective pressure of 5 atmospheres on the air in N C, while a statical pressure of 2·5 atmospheres will restore equilibrium to the column of water R A, which is 25 mètres high.

The same conditions and position of valves, and other parts, are observed in the actual machine, Fig. 41. The valve A is the valve of admission, and is placed in an enlarged portion of the tube. It consists of a zinc cylinder moving in a larger cylinder perforated with holes, F F, which the valve alternately opens and shuts. Its upper surface is made of a conical form in order to allow the water to flow through the holes F F without being thrown into eddies when the valve is opened. The valve rests on a seat surmounted by guides, and is fixed firmly to the side of the tube. This seat is provided with india-rubber buffers for deadening the shock of the valve cylinder should it happen to fall suddenly. The conical cap is fixed to the seat, and not to the cylinder, of the valve, and acts only when the apertures F F are opened. The valve cylinder and seat are constantly bathed in water, so that the apertures can be closed without the valve being subjected to the pressure of 25 mètres of water.

The valve A strikes with great force against its seat, since it is necessary it should open quickly in order that the full force of the water may be applied at once. Although the weight of the valve cylinder is considerable, compressed air is made to act on the upper surface of the piston P, connected with the valve cylinder, and so to hasten the descent of the cylinder. The valve B acts on a similar principle. The valve C is a disc of copper resting in a hollow; it moves from below upwards, and its motion is guided by a cylindrical groove pierced with holes. The valve is placed at such a height that it just touches the surface of the water when the force of the column of water is expended. The valve D for admitting the atmospheric air is a simple valve opening inwards; it is therefore placed at O; but as the valve B opens after the air is compressed, a partial vacuum is formed in the upper part of the tube, which retards the flow of the water until the upper surface of the liquid has reached the point O. A clapper has therefore been placed at the upper part of this tube, immediately below the valve C.

The reservoir of compressed air R′ is an iron cylinder with rounded ends. From its lower part proceeds the tube M leading to the manometer, which exerts a pressure corresponding to that of a column of water 50 mètres high. The tube can be closed by means of a cock, which, however, is always left open when the machine is working. Above the reservoir is a small dome, into which are fixed the tubes for conveying the air from or to the reservoir. The inlet tube is closed by a safety-valve, so as to prevent the recoil of the mass of air in the reservoir should an accident occur.

The machine works in the following manner. At N N is an air engine which acts on the arbor S, on which are mounted the cams, and these, by means of levers, act on the valves A and B in the order above indicated. To start the machine the air engine is first set to work; it lifts the valves and presses on the piston P, so as to hasten the fall of the valve A. The reservoir R′ becomes filled with compressed air. The cock of the manometer tube is then opened. The water of the small reservoir 50 mètres above rushes into the air reservoir R′ until the pressure of the air in the cylinder counterbalances the weight of the column of water. The air first contained in the reservoir is thus reduced to one-sixth of its former bulk, and exerts an effective pressure of 5 atmospheres. The manometric reservoir has a sufficiently large superficial area to prevent its level being sensibly lowered by the action of the other compressors.

The compressed air is next made to work the air engine, and this again acts on the cams. The compressors then commence performing their function of forcing air in large quantities into the reservoir R′. The water which passed from the manometric basin into the reservoir is forced back again. The level of the water in the basin is kept at a constant height by means of a waste pipe. An indicator tube placed against the reservoir R′ shows the height of the water in the reservoir. Should the reservoir become full the machine is stopped.

The cushion of water left at the lower part of the tube at each stroke of the valve is necessary, as it prevents the tumultuous fall of the water, deadens the force of the impact, and causes the column of water to ascend the small limb of the syphon with great regularity. The valve C is raised smoothly, and the small quantity of surplus water which passes each time the valve opens falls into the tube Q, where it accumulates until it reaches a certain height, when it requires to be emptied.

The tube of the compressor has a diameter of ·62 mètre, and the air reservoir a height of 4·05 mètres. The volume of air compressed at each stroke of the ram is 1·223 cubic mètres (43 cubic feet). The compressors make three strokes per minute; each compressor will therefore daily compress 5,283 cubic mètres (186,490 cubic feet) of air into 880·50 cubic mètres (31,081 cubic feet); and the ten compressors at each end of the tunnel 52,830 cubic mètres into 880·5 cubic mètres. The rate of three strokes a minute is slow, but is adopted as being within the limits of safety. It might probably be increased to four strokes a minute without risk of danger. If the increased rate were employed 70,445 cubic mètres (2,486,708 cubic feet) of air could be compressed into 11,740 cubic mètres (414,422 cubic feet) every 24 hours.

The air is also compressed by means of a pump, or by direct action. In the theoretical diagram, Fig. 42, is represented a bent tube having two equal vertical branches provided at their upper ends with the valves A A′, B B′; the first is for the admission of the outer air, and the second for the passage of air into the reservoir R. In the horizontal branch of the tube is a piston which oscillates between the points D D′. This piston is immersed in the column of water which partially fills the vertical branches when the piston occupies a central position, as shown in the figure.

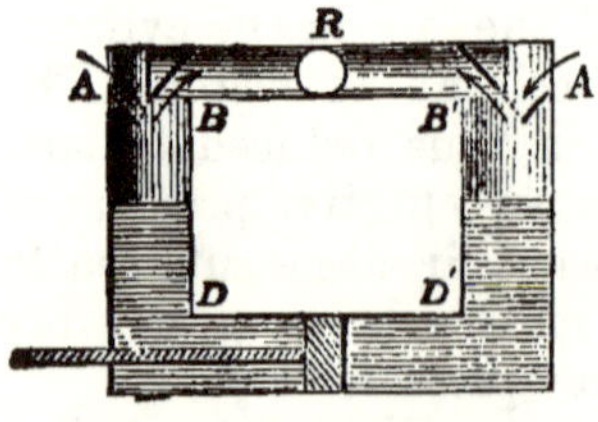

Fig. 42. Theoretical diagram of pump compressor.

If the piston be moved towards the point D the level of the water in the right-hand column will descend to D′. The valve A′ will open, and the tube will fill with air to the point D′. The valve B′ will be closed, owing to the compression of air in the reservoir. On the piston returning towards D′ the valve A will open, and the valve B in the left-hand tube will close. The air in the right-hand tube will be compressed by the water, the level of which has been raised, the valve A′ will close, and the valve B′ will open and give a passage to the compressed air which will rush into the reservoir R.

In short, every time the water is depressed a quantity of air rushes in through the valves A and A′, and every time it is raised compressed air is delivered into the reservoir R immediately the valves B and B′ are lifted. This theoretical diagram will facilitate the understanding of the machine itself, shown in Fig. 43. The piston P is worked by the rod P M, moved by a connecting rod fixed on a waterwheel. The admission valve A is furnished with a disc resting on a ridge, and is kept in position by rods running in guides, and drawn upwards by weights. The orifice of the valve is immersed in a cone constantly filled with water. The upper part during the working is submerged in water about 2 centimètres deep, supplied by a tube which enters the space E. The water filters through a metallic grating at T. A small quantity of it is introduced at each stroke of the piston, and replaces that which is driven in along with the compressed air. Special arrangements are made for getting rid of the water that accumulates in the air chamber.

The piston makes eight oscillations per minute. It has a diameter of ·57 mètre, and a stroke of 1·20 mètre. As the machine is a double-acting one ·61 cubic mètre of air is compressed at each oscillation, or 4·88 cubic mètres per minute, or 7,027 cubic mètres in the twenty-four hours; that is, a result nearly as great as with the other kind of compressors when working four strokes per minute. Six pump compressors, then, will compress 42,162 cubic mètres (1,498,338 cubic feet) into a volume of 7,027 cubic mètres (248,053 cubic feet).

Of these two machines the pump compressor is the simpler and more economical.

In order to show the special application of the power thus obtained to tunnelling we must briefly notice the boring machine. Each machine is fed independently, works independently, and imparts three kinds of motion to the borers;

(1) a strong, rapid, forward motion; (2) a rotatory motion:

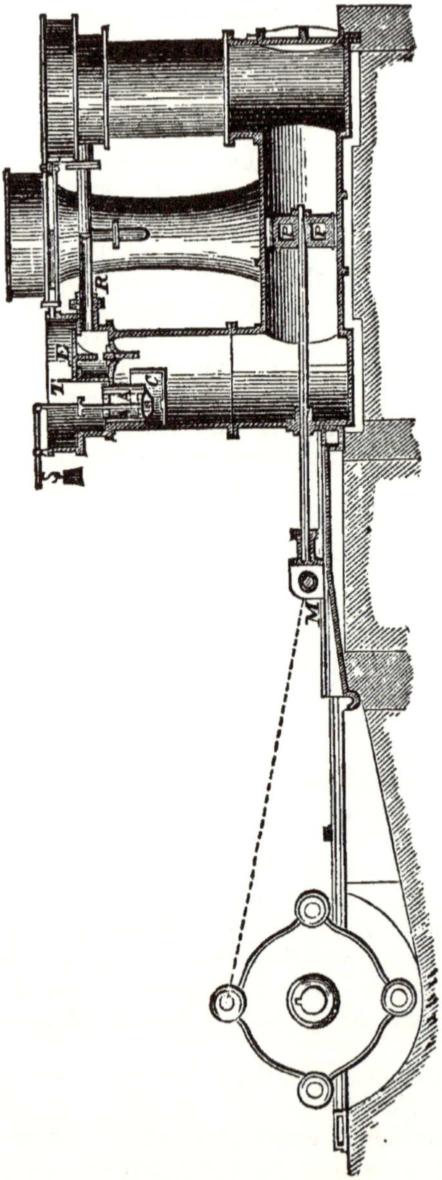

Fig. 43. Pump Compressor.

and (3) a slow advancing motion, which keeps it up to its

work. Fig. 44 is a longitudinal section of the essential parts of the apparatus. It consists of a tube, T T, supplied with compressed air, which it delivers to the body of the pump in which the piston P oscillates. The piston works the tool by means of the arm K M, which carries the tool, and imparts the stroke to it. This pump barrel is mounted on two slides, which embrace the two rods L L, forming the base of the apparatus, and supporting the tool carrier in the collar M. When in motion the pump barrel acts on the surrounding parts marked by the letters C, I, E, H, R, F, F′, U, which are all concerned in promoting the backward and forward movement of the piston and boring tool. All these parts are moved by wheels mounted on the square bar A A, and capable of sliding on it. The rod A acts an important part in regulating the motion of the machine, and is itself moved by the air engine X.

The piston P moves with friction in the pump barrel between the points s s′, and along a groove in the square rod B B. It is pressed at one end by the surface s s, which encircles B, and at the other by the annular surface s′ s′. The surface s s is in communication with the open air when the apertures o′ o′ coincide, and with the compressed air when o o coincide. s′s′ is in constant communication with the compressed air. When the openings o o coincide the piston flies forward, owing to the difference of pressure on the two surfaces; and when o′ o′ coincide the piston flies back. The pump is moved forward by means of the rack K, and the tooth U′ on the rod U. The detent is released by relieving it from the pressure of the spring N. The machine is then pushed forward by means of the projections at E, the shaft B, and the wheel H. When the detent enters the next tooth the forward movement is arrested. The distance between two teeth is ·04 mètre. The rotative movement is effected by the square bar B, which is itself acted on by A by means of a finger on the eccentric I. This finger acts on the wheel R, which turns sixteen times for each revolution of the shaft A. To hasten the backward movement of the tool a wheel, F, is placed on the shaft B, and a wheel, F″, on the shaft A. The motion of the wheel F is transmitted to the wheel F″ by means of the wheel F′. The regulating rod A is moved by the pump piston X. The piston rod acts on the lever Z, which moves in the slide G. From thence motion is imparted through the cog wheels. The action of the entire machine is regulated by the fly-wheel V. The tool, or borer, pierces holes about ·04 mètre (1·55 inches) in

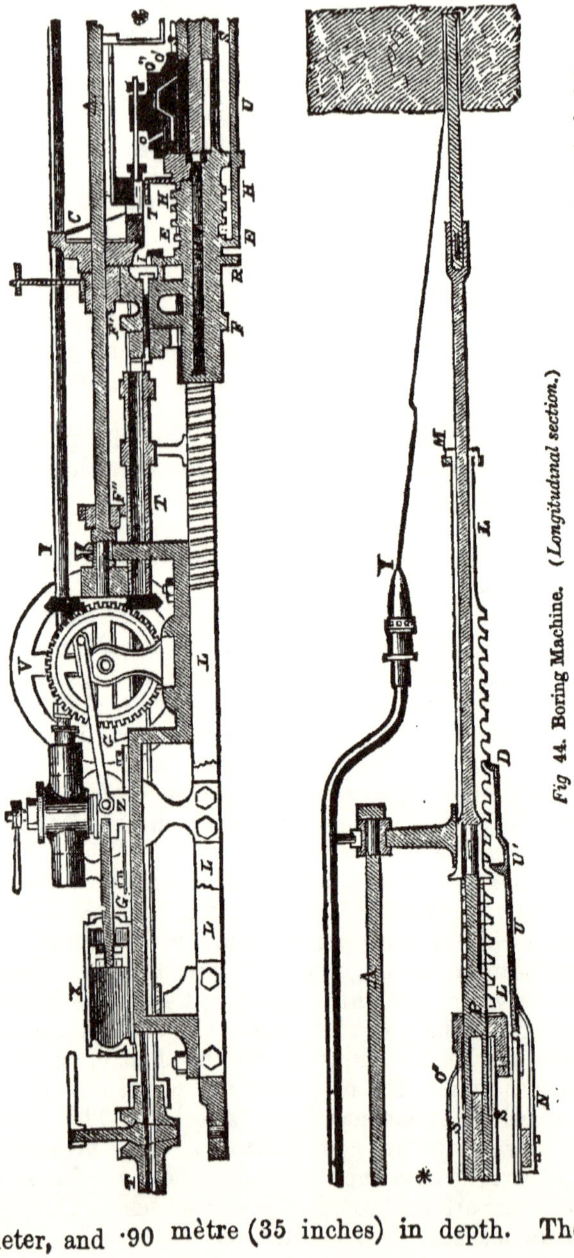

Fig 44. Boring Machine. (*Longitudinal section.*)

[The section being too long for the page, is divided into two parts, the points of junction being indicated by * * in the top and bottom halves.]

diameter, and ·90 mètre (35 inches) in depth. The borers

vary in length from ·50 to 2 mètres. The hole is moistened by means of the small tube Y Y. The daily rate at which the tunnelling at Mont Cenis proceeds is about 1·40 mètres (54·60 inches), but this must vary with the hardness of the rock.

The machine is mounted on a carriage and supported by four wheels running on rails, and moved backwards and forwards as required by an air engine. Each stage carries a number of borers, so that about eighty holes are bored in a space of about 12 square mètres in about six hours, each hole being from ·09 to ·04 mètre wide, and ·90 mètre deep. The holes are cleared out by means of compressed air, and are then filled with charges of gunpowder, which are exploded in successive groups, those in the centre first, and then the others in sets of eight. After each explosion small waggons are drawn up and filled with the débris, which is conducted to the places of deposit, and received by larger waggons. The operation of exploding the gunpowder and removing the débris occupies about four hours.*

CHAPTER XIII.

THE APPLICATION OF WATER-POWER TO GRIND CORN.—THE PROGRESS OF CORN-MILLS, AND DEVELOPMENT OF THEIR MACHINERY.

In applying water-power to grind corn, the usual mode of computing the mill's capability is to ascertain the height and volume of the fall, and thence the available force, expressed in horse-power; and the wheel is so proportioned that when driven at the rate determined, the buckets may be about two-thirds filled by the average supply of water; in order to avoid waste at ordinary times, and also that during the excess in floods, an additional load in the buckets may compensate for the resistance of tail water.

* For fuller information on this subject, and for details respecting the method of ventilation adopted, the reader is referred to a paper in the *Annales des Ponts et Chaussees*, 4e Series, tome v. 1863, entitled *Sur la Percement du Grand Tunnel des Alpes*. Par M. Conte. This description was originally written for the 2nd edition of Tomlinson's Cyclopedia, but is reprinted here by permission of the publishers.

It has been shown in the preceding pages that 12 cubic feet of water per second, or 750 pounds weight, are equal to one available horse-power for each foot in height of their fall when acting by gravity on a well-constructed water wheel; that is to say, the useful mechanical effect is 73 per cent. Mr. Rennie in some cases obtained 80 per cent.; and few wheels and mills, executed with ordinary care and skill, realise less than two-thirds, or 66 per cent. of the waterfall's power.

As the force required to drive machinery of all kinds is now generally expressed in horse-power, except it be immediately applied to work pumps, it has not of late been thought necessary to make direct experiments to show the weight of water required to drive modern mills; their owners are satisfied in knowing or in assuming that they do their work with a given horse-power.

Consequently there are few recorded trials worth notice, since those of Mr. Thomas Fenwick, who found that 300 pounds of water falling 210 feet in a minute, would grind one boll (or two bushels) of corn in an hour.—He says—"The quantity of water expended on the wheel was measured with great exactness; the corn used was in a medium state of dryness; the mills in all their parts were in a medium working state, the mill-stones, making from 90 to 100 revolutions per minute, were from 4½ to 5 feet in diameter."

Therefore it took $\frac{300 \times 210}{2} = 31{,}500$, or very nearly one horse-power to grind a bushel of corn.

To ascertain the friction of a mill when going with velocity sufficient to grind two bolls of corn (or four bushels) per hour, he made a series of experiments in the following manner:—"The mill was made quite clear of corn, and the upper mill-stone raised, so that it would touch as little as possible on the under stone, in its revolutions, then such a quantity of water was admitted to flow on the water wheel (an overshot wheel), as to give the mill, when empty, the same velocity it had when grinding corn at the rate of two bolls per hour, which quantity of water was sufficient to raise a load of 300 pounds with a velocity of 100 feet per minute."

This he therefore considered as the measure of the friction, and concluded, that as the power requisite to grind two bolls of corn per hour, including the friction of the mill, was equal to that which can raise a weight of 300 pounds with a velocity of 350 feet per minute, the difference of the two, which is 300 pounds raised with the velocity of 250 feet per minute, is equal to the power employed in the actual grinding of the corn. This gives a proportion of ·714 to 1, or somewhat over 71 per cent. of useful effect.

Mr. Fenwick, who was of an ancient family in Northumberland, was by profession a "colliery viewer;" he had a high reputation for mechanical and engineering skill, and was living in the schooldays of the author; he wrote several works on practical mechanics and mining, which were at that time esteemed as good authority and went through several editions.

He does not state what kind of mill-stones were employed, but probably they were the "Dutch blue stones" then in use, a kind of lava rock, wrought chiefly in the German quarries near Andernach and Coblentz, brought down the Rhine and shipped to this country from Holland. They had superseded the mill-stone grit, or "grey stones," and have been in their turn displaced by the "French burr stones," now generally employed in grinding wheat. The blue stones

Fig. 45. The fixed cross or rynd.

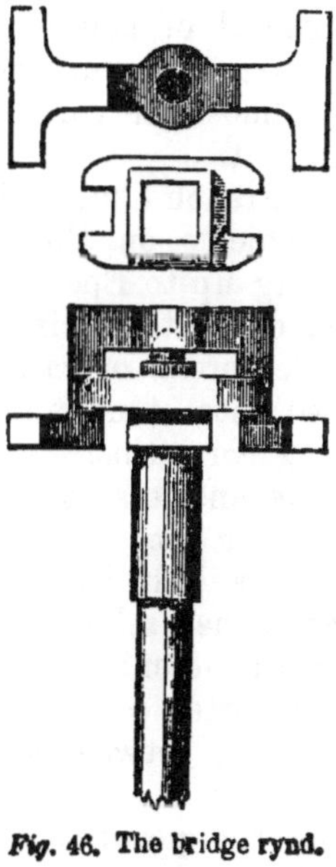

Fig. 46. The bridge rynd.

are but little used, and the grey stones are seldom seen, except in the north for grinding oatmeal, rye, or barleymeal. These blue stones were each in one piece, and were then dressed slightly conical on the face, the lower mill-stone being convex about three quarters of an inch, and the upper stone about an inch concave. The upper end of the spindle was square, and upon it was fixed a wrought-iron cross or rynd, which was let into the upper millstone, so that great exactness was requisite to make the stone true to the spindle. (See Fig. 45.) Afterwards a pivot or centre-point, tipped with steel, was formed on the top of the spindle above the square, and projected up into the eye of the upper stone for about half its thickness; a bridge of iron was let into the stone and rested upon the point of the spindle, a cavity being formed to receive it, and the upper stone was balanced upon the spindle point, at or about its centre of gravity; a driver was fitted on the square with two claws at each end loosely clipping the bridge; the stone vibrated freely like a compass card, resting on the bridged rynd and turned by the driver, so that when the stone was set in rapid motion the centrifugal force caused it to revolve in a level plane, and neutralised any little inequality in the balancing. (See Fig. 46.)

Then it was found better to dress the stones to a level face instead of making them conical, and the upper stone, thus balanced, spun round, and seemed to float, as it were, almost in contact with the nether stone, but not touching it.

The French burr stone is a sharp, porous sandstone, found chiefly on the banks of the Marne, the quarries extending up to Epernay, whence it is brought in boats, and carried down the Seine for shipment at Rouen or Havre; it resembles bread or rather paste in a state of fermentation, changed into stone, and is found so much more effective for grinding corn, that less surface and consequently a smaller millstone answers the purpose. The blue stones were often six feet or more in diameter, and seldom less than five, four and a half feet being considered a small stone. The French stones seldom exceed four and a half feet, or four feet in diameter; consequently they run faster and produce flour of better colour and quality, as they do not retain the corn so long between them, nor rub down the bran with the flour.

From the peculiar formation of the French burr, it is difficult to get a whole stone of uniform quality throughout, and therefore it has been found expedient to build the millstone of selected pieces cemented together with plaster of

Paris, and hooped with iron; the stone is then finished smooth with cement on the back, and dressed to a true and level face, the work or races are cut in, which cross each other when the millstones are in action (see Fig. 47), and the face of the stone having been proved with a paint or trying staff, a strong ruler of hard wood, the stone is sharpened by dressing it all over the face with a broad steel chisel or bill, hatching the stones in parallel lines about an eighth of an inch apart, and in some mills much closer, like the teeth of a file.

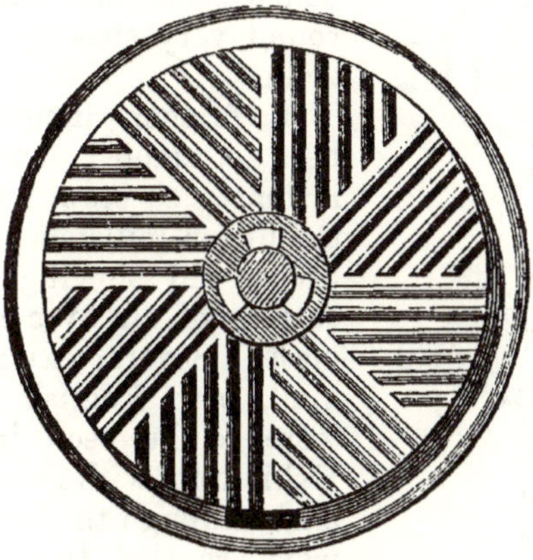

Fig. 47. The Nether Millstone.

To dress or sharpen the millstones with the chisel or bill requires great precision of eye and hand, and much practice to do this well, when the work is close and fine. Those who took an interest in the mill machinery of France in the Great Exhibition, generally found an opportunity to examine, and many to make trial of, a very ingenious tool, or rather a piece of mechanism, somewhat like a ruling machine or dividing engine combined with the mill chisel, to guide the hand of the miller when he dresses the millstones, and enabling him to put in the work, however minute, in parallel lines of equal width. The shaft which holds the chisel works upon a pivot like a small forge hammer, and on this pivot is a screw by which the divisions or spaces between the cuts are determined at pleasure, and the bearings of this pivot, which is about 15 inches long, are $7\frac{1}{2}$ inches apart; these slide on two guide bars and keep the cuts parallel; there is a rest for the arm of the miller, who keeps the chisel going, making the usual sharp cracking noise, but doing his work much more rapidly and regularly with the help of this implement than he could before, besides which, it renders any miller of ordinary skill capable of doing first-rate work.

Lately another mode of hanging the millstones, instead of the bridge or centre rynd, has been introduced: a pair of "*gymbal rings*" in a compact form are fitted to the spindle-head, and the stone is suspended in the same way as the mariner's compass, so that it vibrates equally on all sides without being checked, even by the driver, and as it can be nearly balanced in this way, it swings round very steadily when in full work.

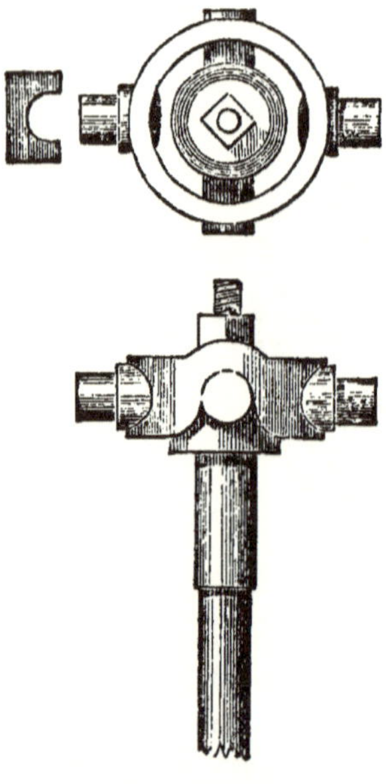

Fig. 48. Gymbal Rings.

A further improvement has been attempted, and in several instances successfully, of making the upper millstones with hollow backs, and blowing into them a current of air by means of a fan blast. Several examples of this kind, both English and foreign, were to be seen in the Great Exhibition, with holes or openings made through the stone in various ways, so that the air may pass between the millstones when grinding, and being thrown out at the circumference or skirt of the stone it assists the grinding and keeps the flour cool.

While so many inventions have been applied to the upper millstone, the lower stone has not been neglected; long after the author was acquainted with mill-work, the nether millstone rested on a framing of wood, and the only means of making the face of it perfectly level or at right angles to the axis of the spindle, was by wooden wedges and packing—a tedious and difficult operation. This was obviated by a cast-iron seat with the means of adjustment by set screws, and lately a wrought-iron cradle of an equilateral triangular form receives the nether millstone. The angles being equal, and a screw being placed at each angle, the millstone rests on three equidistant points, and is in this way as easily set as a spirit-level. This seems a favourite plan with the American and the French mill-wrights, but although the three points must bear equally, many of the English mechanics prefer adjustment by four points, by which they obtain two level lines intersecting each other.

The stone being truly levelled on the face, is now often adjusted and fixed horizontally also by three or four screws dividing the circumference; but in many mills it is retained by a strong curb of wood surrounding the stone, and rising up within an inch or two of its face. This curb serves also as a base for the wooden case enclosing the millstone, and through it the meal is discharged by the spout below into sacks or binns as may be most convenient. In several of the modern mills the case is made sometimes of galvanised sheet iron, and the meal descending by the spout falls into a long trough fitted with an endless screw of a coarse pitch and deep thread, sometimes of cast-iron, and sometimes of thin plate of equal length with the trough, revolving in it, which conveys the meal to some convenient point, whence it is carried up into a higher loft by means of an endless belt or web fitted with light iron or tin buckets, and called by the millers a "Jacob's ladder;" it is then prepared for the dressing-machine.

The most conspicuous adjuncts of the millstones were the hopper, with the feeding trunk or shoot, to bring the corn into the hopper from the floor above; the shoe to carry it over the eye of the millstone; the mill clack or "damsel," to shake the shoe and supply the corn evenly and regularly from the hopper to the mill; the cord and peg to raise or lower the point of the shoe and augment the feed; the wooden spring to make the shoe recoil; and the warning bell to call the negligent miller to his empty hopper. Pleasant it was to see all this simple, but ingenious apparatus, steadily and quietly, yet with an appearance of life, instinct, and good will, cheerfully perform its task to the sound of the merry mill clack. The hopper and its attendants are disappearing in modern mills, and a brass receptacle somewhat resembling a Roman urn in form has superseded it.

The iron apparatus for adjusting and securing the millstones, involved an iron frame and columns to carry it, instead of wood, and hence many mills of recent date have been built fire-proof. In such cases stone foundations and flooring have become requisite, and the mill-spindles, instead of being stepped upon the middle of the foot-bridge (a lever of wood fixed on a joint at one end, and regulated at the other, between the timber uprights carrying the millstone floor), has been brought down to a cast-iron pedestal resting on the masonry, and containing within it a compact combination of wheels or levers and screws, by which the space between the millstones may be altered at pleasure.

It has been proposed to use annular millstones by removing the central part, where little work is done, and replacing it with a plate of iron, leaving the skirt, or as it were a ring of stone; no practical examples of this kind have come under the author's observation, except so far as one of the conical corn-mills partakes of this character; but this must be described separately.

There are three modes of driving millstones.

The first was suggested by the use of the undershot water-wheel, which, revolving rapidly, required but one increase of speed, and one change of motion from the horizontal shaft to the vertical spindle, and consequently only one pair of wheels, namely, the face-wheel or pit-wheel on the water-wheel axis, and the lantern-wheel or trundle on the mill-spindle; all the machinery being made of wood except the pivots of the axle and the mill-spindle.

This simple, but efficient arrangement, may still be useful in remote settlements, where neither capital nor labour are abundant. It was the usual plan of a mill during the greater part of the last century, and the works of most writers on mechanics and mill-work of the time contain tables showing the number of cogs in the wheel, and staves or rounds in the trundle, to drive a 6-foot millstone sixty revolutions per minute. The cogs and rounds of wood have given place to bevelled wheels of cast-iron, but the principle remains the same. By this means any number of millstones may be placed in a straight line, or in two lines parallel to each other, each millstone requiring a pair of bevelled wheels, one of the wheels (the largest) having wooden teeth, that they may work more smoothly and with less noise. Mr. Smeaton used this plan for a line of millstones in 1781.

The second method came in with the overshot-wheel, which, going at a slower rate, required the speed to be increased twice between the water-wheel axis and the millstone. So the pit-wheel turned a bevelled pinion fixed on an upright shaft, which formed as it were the centre of the system; on this was also fixed a spur-wheel of large diameter, say from 6 to 10 feet, and round this were ranged in a circle the millstones varying in number with the power of the mill, seldom exceeding six pairs, although small mills had only two

These were driven by the large spur-wheel, by a spur pinion

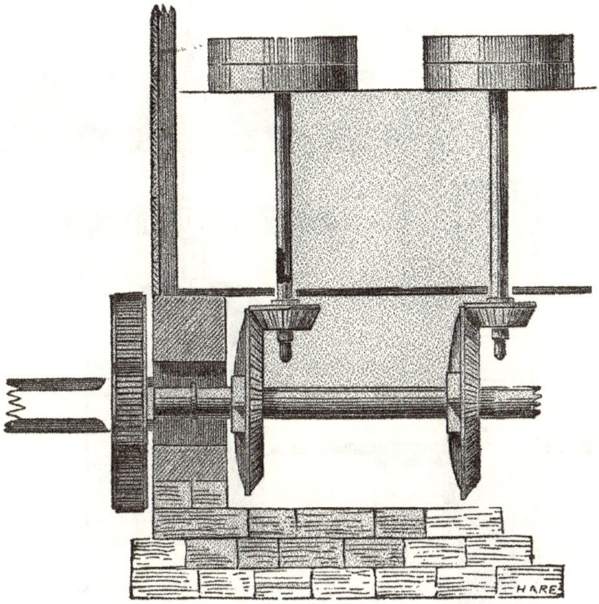

Plan.

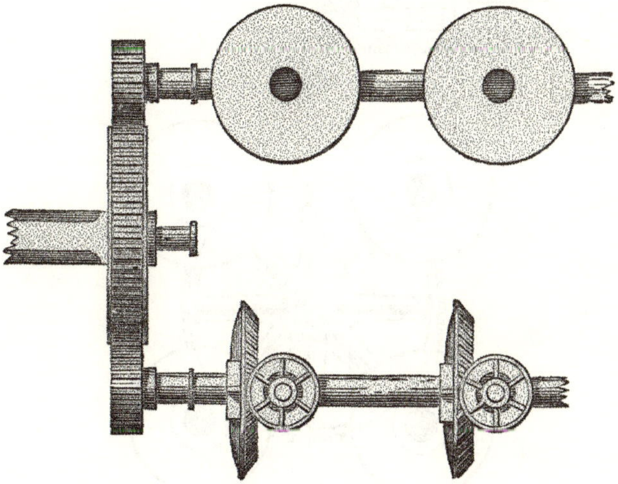

Fig. 49. Method of driving wheels.

upon the spindle of each pair of stones—any one of which

could be thrown out of gear and kept still while the others

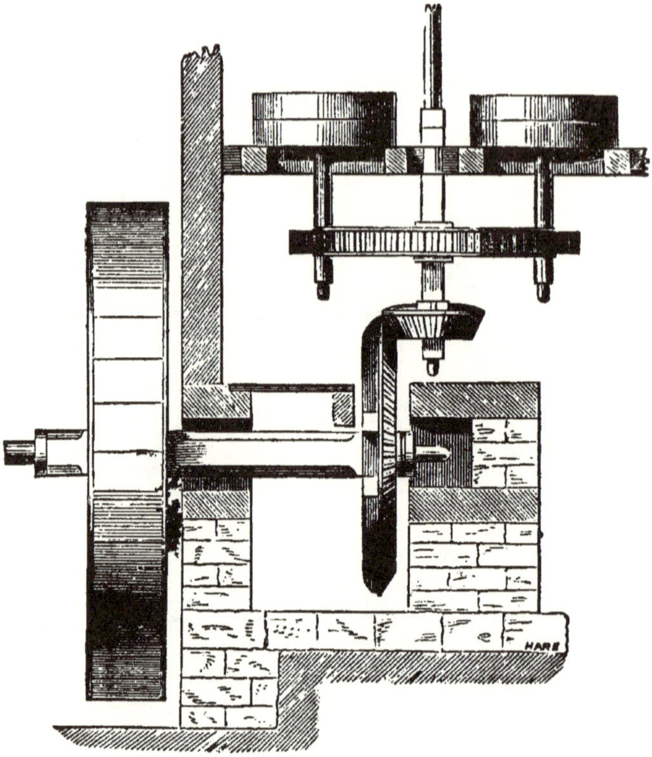

Plan.

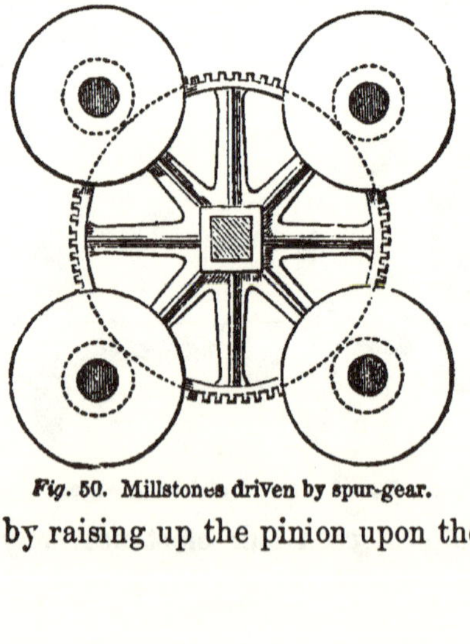

Fig. 50. Millstones driven by spur-gear.

were at work, by raising up the pinion upon the spindle clear of the spur-wheel, and retaining it in that position by a screw. In the older mills a pair of stones were kept at rest by taking a couple of staves out of the trundle.

The third mode of driving the millstones is by belts; the author first saw it used in a large flour-mill at Attercliffe, near Sheffield. The inconvenience attending this mode consists in the necessity of placing the belts at different levels when the millstones are placed round a vertical driving-shaft, and in the difficulty of disengaging them when so placed, so as to set any one pair of stones at rest while the others work; and when a line or two lines of millstones are driven from a horizontal shaft the belts are twisted in passing to the vertical spindles, whence there arises a number of practical dif-

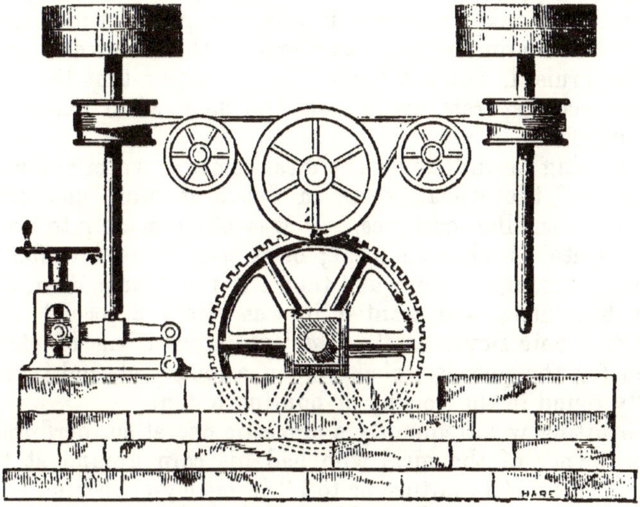

Fig. 51. Millstones driven by Belts.

ficulties—in keeping the belt in its proper place when at work and when at rest, and also in keeping it at a proper tension, so that the strain upon the belt and spindle may be equal, sufficient, and not in excess. The advantages are supposed to consist in the saving of a pair of bevelled wheels to each pair of stones; but as there are three pulleys and a belt substituted for them, the economy is rather doubtful. There must be one broad pulley on the horizontal shaft, and two on the spindle, the one fast and the other loose, or *vice versâ*, besides the requisite apparatus for giving the belt its proper tension, the means to prevent the neck of the spindle from being drawn to one side by the pull of the belt.

In preparing linseed for the crushing action of the "edge-stones," as used in oil-mills, it is found necessary to bruise the seed between two rollers, like those of a malt mill, otherwise this edge-stone would be much longer in doing its office, and most of the seed would escape without being crushed at all. Acting on these observations, the French millers have lately adopted rollers to bruise their wheat before it is delivered to the millstones. The bruised corn is much more readily converted into meal than when delivered whole into the millstones, and it is said that by using these rollers both time and money are saved. Since 1851 these rollers have found their way into some of the English mills. In some of these machines there are two pairs of rollers, one above the other; the upper pair bruise the larger grains of corn, and the lower pair take those which pass through the wider space above: the reason for this exactness is, that the wheat may only be bruised, but not broken in pieces, so that the skin or bran may separate easily from the flour when under the millstones.

On leaving the millstones the meal comes down mixed with the skin of the wheat, some in flakes as bran, and some reduced to smaller particles. It was often made into bread in this state, as wheaten meal; or it was thought sufficient to pass it through a sieve and take out the coarse bran, and when fine flour was wanted it was sifted a second time through a hair sieve, or "*bolted*" through a bag of cloth, woven for the purpose, placed upon a long reel, and turned rapidly round in the bolting or bunting hutch.

The sifting was frequently a domestic operation, performed by the owner of the meal who had his corn ground at the mill, and paid "a multure or toll" in money or in kind; the miller who ground for hire being prohibited by Act of Parliament from buying corn to grind for sale.

Subsequently woven wire was used instead of hair-cloth, and the use of a brush to clear the meshes of the fine sieve, soon introduced *the dressing-mill*, the invention of which was easy, as *the bolting-mill* was already in use.

The reel being fitted with brushes revolved at a high speed in a cylindrical frame or open-work case of wood, 6 or 7 feet long, and 18 or 20 inches diameter, lined with wire-cloth, and placed in an inclined position, the finest wire at the upper end; the size of the wire and meshes increased towards the lower end, where the bran was delivered. The cylinder was either stationary, except when turned round in its bearings by hand, or it had (as an improvement) a slow rotary motion, to bring all parts of the circumference to the top in succession, to keep the wires clear. The motion of the brushes could be reversed, so that by revolving in opposite directions they might wear equally on both sides, and the bristles be prevented from taking a permanent set in one direction; adjusting screws were inserted in the body or axis of the reel to give them the proper pressure against the wires, or to advance the brushes as they wore down. The meal was taken into the head of the cylinder by a "shoe" or spout of wood covered with leather, which was shaken by cams on the spindle, and the enclosed cylinder rested over a series of hoppers to receive the flour of each degree of fineness, as it was driven through the wire-work by the rapidly revolving brushes, generally eight in number. Their speed has been increased since the author first examined these machines, from 320 revolutions per minute to 450; some run 500, and even more, the velocity having of late been much augmented, and the fineness of the wire-cloth also. The wire-work at the upper end of the cylinder being now generally *No.* 54, that is to say 54 wires in a lineal inch, or 54 × 54, say 2,916 meshes in a square inch, woven with wire of 34 gauge; some machines have wire-cloth as fine as *No.* 84, woven with wire of 38 gauge, or 84 wires in the inch, making 7,056 meshes in a square inch; and the Lincoln millers have used wire-cloth so fine as *No.* 92. As the wire occupies two-fifths of the space, the aperture of the mesh is very minute, and the flour dressed through it exceedingly fine, but wire-cloth so fine as this is not often used. The wire-work becomes coarser towards the lower end of the cylinder where the bran passes off.

The partitions dividing the hoppers were made movable on hinges at the bottom, so that their mouths might be widened or contracted in proportion to each other, and more or less fine flour be taken out of the meal, or thrown into the next quality as the customer might desire.

The dressing-mill had now reached its climax, so far as principle is concerned, although many improvements in detail afterwards took place; as, for instance, the substitution of iron for wood, in the case or frame of the cylinder, and the size and inclination of the cylinder itself, which has been increased from 20° to 45°, as the speed has been increased, so as to dress the flour more quickly; for the sooner this operation can be completely performed the better is the flour. The application of external brushes has been introduced to prevent the wires from clogging, and it is found to act beneficially. Attempts have been made (some of them with satisfactory results) to place the dressing-mill quite perpendicular; but the flour falls to the bottom as soon as the centrifugal force ceases, unless prevented by means of shelves or partitions, while the inclined cylinder better admits of the convenient arrangement of hoppers for varying the quality of fineness of the flour, so that there does not appear to be much advantage gained by carrying the cylinder's inclination beyond 45°. Many improvements have been made in the manufacture of bolting-cloths, which are now woven without a seam, and the French millers have substituted silk for woollen fabrics to dress their flour, the English millers have also adopted them in many instances.

The corn delivered to the mill by the English farmers is generally well cleaned; but since the importation of foreign corn in large quantities, which in some instances has lain in heaps on the ground exposed to the weather until it could be shipped, it has become requisite in extensive mills, which manufacture flour from all kinds of corn that may be brought to them, to provide machinery for separating light or damaged grain from the sound wheat, and also to free it from sand and stones.

For this purpose, corn-screens and cleansers of various plans, made of wire or perforated metal, aided by a fan-blast, have been the means employed. The system of washing the corn to free it from impurities, and afterwards drying on a kiln, "*la voie humide*," of the French, is not much practised here; the millers finding the opposite plan, or dry system, "*la voie sêche*," more convenient.

Suppose a vertical cylinder or drum, revolving 280 to 350 times in a minute, covered externally with sheet iron, pierced with an infinite number of small holes, so that the asperities of their edges project outwards, and placed inside of another cylinder, lined internally with perforated sheet iron, the sharp edges of the holes projecting inwards with an angular space between them just sufficient to admit a grain of wheat, care being taken that it does not "pearl" or skin the corn. Let the internal drum be about 5 or 6 feet high, and its diameter about 20 inches. On the top of the drum let there be a fan, with four blades or wings, revolving very rapidly, and suppose the corn to be carried up by the Jacob's ladder and dropped into this fan, it is obvious that all light corn, dust, straw, chaff, and such like things, will be driven off by the wind of the fan, and that all heavy bodies, as stones, nails, sticks, and the like, will be driven off by centrifugal force, and may be caught in an enlarged part of the case properly designed to receive them; but such particles of earth and grains of bad wheat as are of the same size and weight as the good corn fall down with it into the space between the two cylinders, and pass in a spiral direction from top to bottom. In passing downwards, they receive a severe rasping from the two plates of iron, which, like nutmeg graters, breaks the skin of the damaged corn, and crack the little pellets of earthy matter or sand in their descent, and also cleanse the corn of sprouts, beards, and other protuberant excrescences. At the bottom is another fan, which separates the dust, smut, and other light matters rubbed off the grain, which, being caught and retained, the corn passes into a slightly inclined cylinder or cribbling machine, about 20 inches in diameter, and 12 or 13 feet long, covered with smooth plates full of small holes, and by it all things smaller than a grain of wheat, but of the same specific gravity, which the two fans have failed to separate, are taken out of the corn,—larger things of equal weight having been grated down by the drum, and heavier things driven off by the fans.

When there is plenty of good water and plenty of room, the humid mode of treatment has its advocates, and they have many good reasons to advance. The author well remembers, many years ago, visiting the mills of Mr. Pilling, at Mirfield, near Bradford, Yorkshire, who very successfully treated his corn in this way. His mode of drying it was peculiar. Suppose a building three stories high; and on the ground floor a stove or cockle, such as is used to dry yarn, consisting of an iron chamber or oven, with a fire-grate in the bottom of it, enclosed in another iron case, leaving a space of 15 or 18 inches between the two on every side, so that fresh air could pass through and be heated: by the side of this cockle was placed a powerful fan-blast to drive the air through this space, which, being heated as it passed, went up through the floor above, and filled a low room or chamber, made air-tight, or nearly so: this room was only about five feet high between the floors. The first floor was laid with Yorkshire pavers on cast-iron joists, and the second floor was laid with perforated kiln plates, pierced with an immense number of small holes cast in them. This was a lofty room with a cowl, like a malt-house, and upon this floor was spread the wheat to be dried, through which the hot air, urged by the action of the fan, found its way.

The arrangement was not only found to be very useful in drying corn after it had been washed, but it was also very beneficial in drying corn that had been exposed to wet on the field, thereby preventing it from spoiling, as well as in giving the requisite degree of dryness which millers prefer. The extreme hardness and dryness of much of the Italian wheat, and the produce of some parts of the south of France, render the corn, unless it be moistened to make it "grind kindly," apt to splinter and break, like ground rice, rather than to flag out into meal, when the flour and bran readily separate.

The corn being treated in the dry or the humid way, if it require either, or if it be clean English wheat carefully dressed and winnowed by the farmer, it is brought to the proper shoots to be delivered into the millstones and ground, by means of the endless screw and the Jacob's ladder, or in mills of less size and pretension it is shot from the sack by the miller as he wants it.

When the wheat arrives at the mill in the barge or the farmer's waggon, it is hoisted up by the sack tackle—a very simple but ingenious machine, which ought to be much more generally used in all places where merchandise is to be lifted and loaded or warehoused.

Let us return to the mill driven by spur gear, with a vertical shaft in the centre, extending from the bottom of the mill to the top. On the upper end of the vertical shaft is fixed a horizontal wheel, bevelled on the face and clad with wood fitted into it endways and turned smooth: against this works another wheel made in the same way (without teeth in either of them), but the grain of the wood being at right angles with the faces of the wheels; the two, when brought in close contact by means of a compound lever, *bite* together with sufficient adhesion to hoist a sack of wheat or flour. This contact is produced by the miller pulling a cord attached to the lever, and the wheels are brought forcibly together; the upright shaft always revolving while the mill is going, but the other wheel, which carries a roller and a chain, revolves only when the cord is pulled, and while the two wheels are kept in contact.

The miller makes a noose with a ring at the end of the chain, tackles the sack, and pulls the cord; the wheels are brought into contact, and the sack ascends, opening the trapdoors in each succeeding floor, which again close as it passes through them, until the sack attains the required height, when he ceases to pull the cord, and lands it. In like manner, when he wants to lower a sack of flour into a barge or waggon, he tackles it with the noose, and lifts it by pulling the cord, then shoves it off, and preventing acceleration by an occasional and dexterous pull of the cord, he drops it into the hands of the bargeman or waggoner, who guides its fall into the proper place.

In 1852 the large flour manufactory known as "The City Flour Mills," and situated near Blackfriars Bridge, commenced operations. The following is a brief description of its machinery:—It has two engines working together on one shaft; they are large "steam-packet" engines, each of 125 horse-power, or, together, 250. There are 60 pairs of millstones, all of them 4 feet in diameter, and making about 128 revolutions per minute; the upper millstones have hollow backs, and a blast is sent into them; thence it passes through the eye and through holes in the top stone; the air quickens the grinding, and makes the stones work cool. The millstones are placed in two rows, one against each side wall, on one floor. The flour is received in covered metal troughs, in which endless screws work the meal along into boxes, to be carried away by the Jacob's ladders or elevators.

The dressing-machines are hexagonal, and are covered with silk; they are 3 feet 4 inches in diameter across the angles, and 34 feet long, and have a slight inclination, about 20 inches in their whole length, and are driven at about 28 or 30 revolutions in a minute. The shafts of these machines are hollow, and also have holes through them; but the lower half of the shaft is closed; a blast is driven into the upper end of the hollow shaft, and blows through the holes into the inside of the dressing-mill, for about half its length, and through the silk, so that it cools the flour, cleans the silk, and quickens the action of dressing.

There are some wire-machines for brushing the bran, but all the flour is dressed through silk, the chief object being to make fine flour. There are, besides, corn-cleansers, also with a blast, somewhat like those already described, and sack-tackles, worked by belts, many of which are double-acting, at every point where they can be made useful, so that manual labour may be economised at every stage of the work by the intervention of mechanical power.

All the millstones, and most of the machines, are driven

by belts, which lessen the noise considerably, and much ingenuity is displayed in their application to prevent lateral stress on the necks of the mill spindles and to obviate the other objections to belts before mentioned: the spindles are long, and have no foot-bridges, each being stepped on a hollow pillar, containing a regulating screw, and the upper stones are hung, like a mariner's compass, on gymbal rings, and carefully balanced: so that altogether it may be said this is a

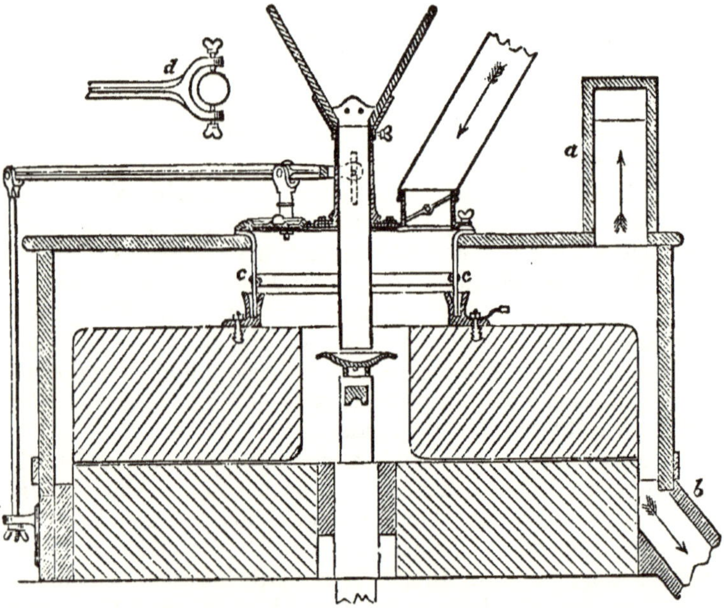

Fig. 52. Method of Ventilating Millstones.

a, The waste air trunk carrying up the dust or stive.
b, The meal spout.
c, Collar or ring of leather to confine the air brought in by a flexible pipe.
d, Forked end of the lever to regulate the supply of corn to the millstones. The arrows show the course of the air.

flour factory rather than a corn-mill; and in order that nothing may be wasted, exhausting-fans are applied to the millstone cases, producing a slight draught of air sufficient to collect and deposit in a proper chamber the dust, or "*stive*," as millers call it; but not strong enough to carry up the meal. This also keeps the mill cleaner, and the stive, which is finer than flour, can be made useful. The author is much indebted to Messrs. Swayne and Bovill, who have designed and erected the machinery of this mill, for their readiness in affording him information and access at all times to this magnificent establishment.

Mr. Bovill proposes to apply blowing and exhausting-fans to ordinary millstones already in use, without making holes through them, by sending the current of air through the millstone eye to be discharged at the circumference, and to collect the dust, or "stive" from the millstone cases by the exhausting-fan, and prevent its flying about the mill, carrying it up through an air-trunk, and discharging it into a chamber of lattice-work lined with bunting, or some similar thin woollen cloth, permitting the air to escape, but retaining the dust. The annexed woodcut shows the proposed arrangements. (See Fig. 52.)

CHAPTER XIV.

THE CONICAL MILLS.

Of the conical corn-mills exhibited in the Crystal Palace, there were two which more particularly claimed attention. One of these was a cone, or rather a conoid of stone of peculiar form, base upwards, which fitted, or nearly so, into a block of stone hollowed out to receive it, and in which it revolved, like the ancient "*quern*," or like the old mill found at Pompeii reversed; for in that antique mill, the cone or conoid stood fixed upon its base and the stone casing revolved, but in this modern adaptation the case is fixed, and the conoid turns.

The form given to this millstone was also recommended by the exhibitor as applicable to the pivots or steps of upright-shafts, screw propellers, turbines, and other mechanism where the revolving surface must resist vertical weight or end thrust.

The inventor proposed to exhaust the air from below the millstones by means of a ventilator or fan, and thus to draw down a current between the stones to keep them cool. There was much ingenuity displayed in various contrivances about this mill and its adjuncts; the trouble and inconvenience occasioned by the cutting and heating of steps and pivots subject to end pressure, even in an ordinary lathe, are well known, and should the mill itself not prove so successful in practice for grinding corn as the inventor expects, it may answer well in grinding painters' colours; the form of step he recommends may in many cases be useful to lessen the friction of machinery. Mr. C. Schiele, of Oldham, near Manchester, is the proprietor of this mill, and the curve he has

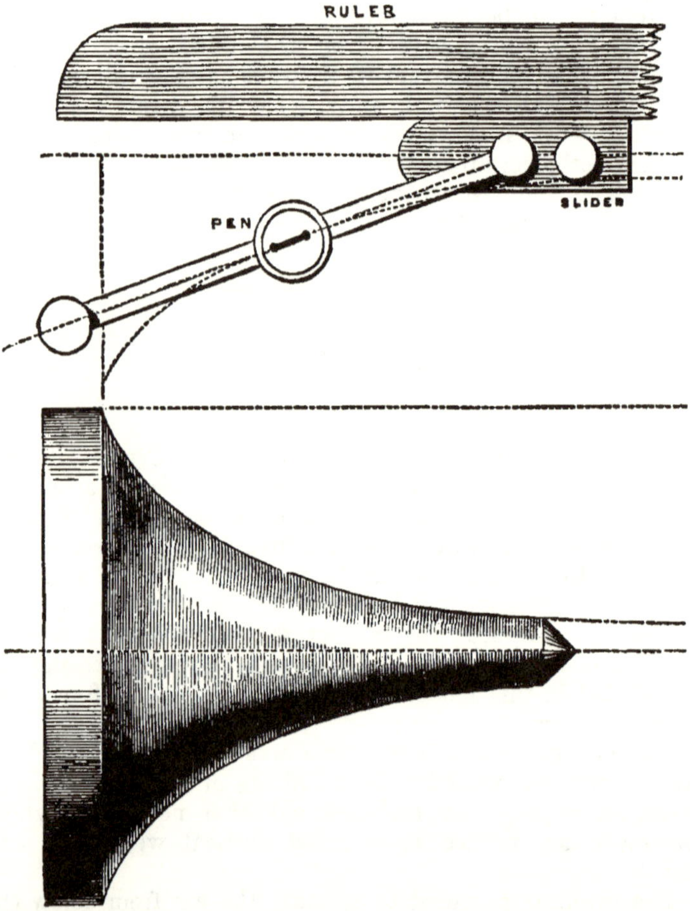

Fig. 53. Schiele's Mill.—Mode of striking the curve to form the millstone or pivot.

adopted is one discovered by Huyghens, in his investigation of the cycloid. It is one of those singular and beautiful curves called "tractories," and in this case it is produced by drawing the centre point of a radius bar along a straight line, which is the axis of the curve. The radius bar carries a pen, the nib of which is in the line of the radius. At the commencement of the curve the radius is at right angles with the axis, but the radius bar, turning freely upon the centre point, as it is drawn along the straight line, or axis, the angle it makes with this line varies continually, becoming more and more acute like the tangent of a catenary or a parabola. Dr. Peacock has shown that the mechanical tractory of a straight line, upon a perfectly smooth plane, is an inverted semi-cycloid; but, in this case, the retardation produced by the pen and paper causes the curve to be infinite, and from its peculiar properties it has been termed "the equitangential tractory." (See Fig. 53.)

The conoid is formed by this curve revolving on its axis.

The other conical flour-mill, which deservedly attracted much notice in the Exhibition, was the invention of Mr. W. Westrup, a practical London miller. It differed entirely from any other flour-mill hitherto used. Each mill, so to speak, has two pairs of millstones combined, working together, the one pair placed above the other, so that the upper pair commences the grinding process, and the lower pair completes it; there is a space between the two pairs of millstones about 27 or 30 inches in height, and the greater portion of this height or space is used as a vertical dressing-mill, the spindle which drives the stones being fitted with brushes, and the space enclosed with a cylindrical screen of fine wirecloth mounted on a frame in the usual way. The upper millstones are fixed, and the lower stones revolve, and both the upper and lower stones are placed upon one spindle. The upper stones are each made in two parts, or semicircles, bolted together, for convenience of fixing and displacing when needful, and they are capable of adjustment by means of fixed wedges or inclined planes, on which they rest, so that by the action of a screw and wheel a partial horizontal turn or twist of either of the upper stones causes it to slide up or down on these bent wedges or inclined planes which are placed round the circumference of the stone. It is thus raised or lowered, and the grinding space adjusted with great facility. The lower millstones, which revolve, are convex, and the upper stones concave and annular; for the stones being of small diameter the eye of the stone is large in proportion. The diameter is about 2 feet 6 inches, and the grinding surface on each side of this ring of stone 8 or 9 inches broad; the rise or bevel of the cone in that width is about 4 inches. The stones being small, necessarily revolve rapidly,

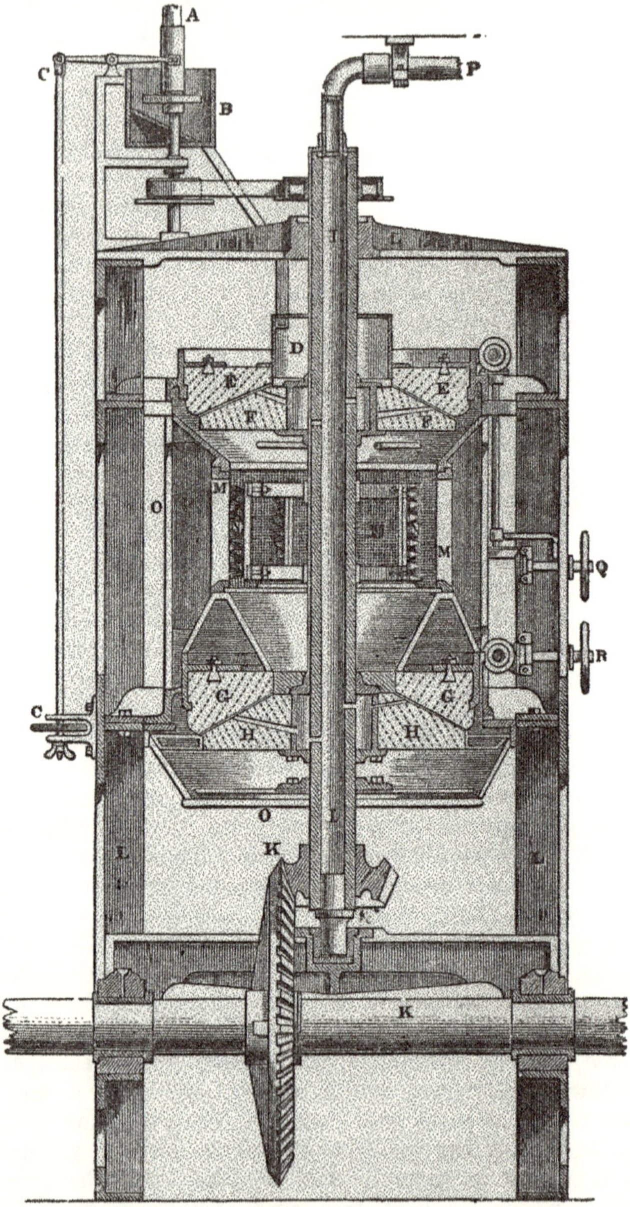

Fig. 54. Westrup's Conical Mill.

References to Mr. Westrup's Conical Mill. (*See Fig.* 54.)

A, Feeding-pipe to supply corn to the millstones.
B, Apparatus to regulate the supply.
C, Regulating lever to adjust the same.
D, Chamber over the eye of the millstone to receive the wheat from the regulator.
E, Top stone in the upper pair of millstones which in this mill is stationary.
F, Nether stone of the upper pair. In this mill it revolves.
G, Top stone (stationary) of the lower pair.
H, Nether stone (runner) of the lower pair.
I, Hollow spindle on which the runners or revolving millstones are hung.
K, Bevelled wheels and driving shaft.
L, Iron framework sustaining the whole machine.
M, Upright wire cylinder acting as a partial dressing machine.
N, Revolving brushes acting against the wire.
M, O, Wooden case enclosing millstones and wire cylinder, to the bottom of which the spout for the meal is affixed.
P, Pipe to convey cold air to the faces of the millstones by means of the hollow spindle.
Q, Regulator for adjusting the upper pair of millstones.
R, Regulator for adjusting the lower pair.

say about 250 revolutions per minute, and the spindle being hollow from the top, a pipe is fitted into it by a swivel-joint, and a blast of air driven by a fan is carried down the spindle, which is closed at the lower end, and distributed through holes in the running stones into the grinding space, so that the meal is immediately blown out, and the grinding surfaces kept clear. The finest flour is brushed through the wire-work of the vertical cylinder, and received in a casing of wood. The larger particles and portions of the corn imperfectly ground pass into the lower pair of stones, and are reduced into meal ready for dressing in the ordinary way.

As by this arrangement of parts the corn cannot be delivered into the centre of the upper millstones, a hopper or chamber is placed on one side, with a sliding tube or feed-pipe in the top of it, and an upright spindle carrying a dish, which revolves quickly, and evenly distributes the corn. This description will probably enable the reader to understand the annexed engraving, which is copied from a section obtained from the inventor. (See Fig. 54.)

The manner in which this mill does its work is very satisfactory. The corn is so short a time in passing through it, that the bran is delivered in large flakes, many of them nearly the entire skin of the wheat, and the grinding being so quickly done the meal comes away comparatively cold. The fine flour driven through the intermediate dressing-case is the heart or kernel of the wheat, and is suited for pastry or confectionery.

There is a vertical dressing-mill separate from the mill-stones, in which the flour is finished for the market, and there has been much ingenuity exercised in contrivances to obviate the inconveniences incident in general to the perpendicular cylinder. The axis carries a series of shelves or tables, which in succession receive the meal and scatter it by their centrifugal action; but it is questionable whether any real advantages are obtained by carrying the inclination of the dressing-mill beyond the angle of 45°, for at that angle, with a high speed for the brushes, the meal must describe a spiral track for several revolutions in traversing the length of the cylinder, and the adaptation of hoppers, with movable partitions before described, is of great practical convenience.

The author hopes that the circulation of this little volume may stimulate inquiry and research among the practical and operative men into whose hands it may fall; and that the facts and circumstances which have rather been indicated than described, may form as it were the text, which in more able and experienced hands may be amplified and illustrated. The work has been hastily written at intervals, when the author could spare a short time to add a few pages or to note down a few observations as they might occur, without much opportunity of arranging them afterwards.

This will be apparent to all who may read this book, and it will also be noticed that the words of the authors quoted throughout the volume are generally given as they were found, in the phrase of their own style and time, so that the lesson may be learnt as they taught it. It has been exceedingly gratifying to the author, and he has much pleasure in acknowledging it, that whenever he has had occasion to verify a fact, to correct a statement, or to ask information, he has in every instance been met with the greatest frankness and candour, and every question asked has been fairly answered—some by letter and some personally. It is difficult to mention names when so many persons have been referred to and all have acted alike, yet he feels he should be ungrateful to the following gentlemen did he not acknowledge the obligation he is under. He has especially to thank Messrs. Ransom & May, of Ipswich, well known as manufacturers of agricultural implements, and of machines for cleansing and dressing wheat; Messrs. Bryan Corcoran, and Co., of Mark Lane, importers and makers of millstones and grinding machinery; Messrs. W. Mountain & Sons, of Newcastle-on-Tyne, celebrated for their wire-work and dressing-mills; Messrs. Swayne &

Bovill, the millwrights of the City Flour Mills; Mr. Westrup the inventor, and Mr. Middleton the manufacturer, of the conical corn-mill; Sir W. Armstrong, the inventor of the hydraulic crane, and maker of the water-pressure engine; Messrs. Easton & Amos, manufacturers of the improved water-rams; and many other gentlemen.

In writing this treatise, the works of Smeaton, Hutton, Sir Robert Kane, and Mr. Beardmore, and the papers of Mr. Blackwell, Mr. Rennie, MM. Morin, Poncelet, Piobert, Colladon, and other foreign engineers, have been quoted, and the Transactions of the Royal Society, the Institution of Civil Engineers, the Society of Arts, the British Association, the Professional Papers of the Royal Engineers, and other records consulted. In all of those above-named the reader will find much information and many valuable experiments in detail; but to do justice to the subject which forms the concluding part of this little book, *The Progress of Corn-mills, and the Development of their Machinery*, would require a large and amply illustrated volume, rather than a simple rudimentary treatise on the means of preparing the staff of life for the support of man.

APPENDIX.

THEORY OF THE CENTRIFUGAL PUMP.

OF the total work employed in producing rotation, that portion which represents the force in the normal or "the centrifugal force" is that alone which under any circumstances can become "duty" in the centrifugal pump.

Generally—

Let G = the weight of a revolving body; and hence its mass $M = \frac{G}{g}$, g having the usual relation to gravity.

r = the radius of revolution.

v = the velocity of revolution in the circumference.

P = the force in the normal, or the centrifugal force.

$$\text{Then } P = \frac{M v^2}{r} = \frac{G v^2}{g r} = 2 \frac{v^2}{2g} \frac{G}{r}; \text{ hence}$$

$$P : G :: 2 \frac{v^2}{2g} : r.$$

That is, the centrifugal force is to the weight of the body in revolution as twice the height due to its velocity is to the radius of revolution.

The resistance being uniform, we may express v in terms of the time of revolution T with the radius r, and

$$P = \left(\frac{2 \pi r}{T}\right)^2 \frac{M}{r} = \frac{4 \pi^2}{T^2} M r = \frac{4}{g T^2} G r;$$

and as the constant $4 \pi^2 = 39{\cdot}4784$,

$$P = \frac{39{\cdot}4784}{T^2} M r = 1{\cdot}224 \times \frac{G r}{T^2}$$

or if n = the number of revolutions per minute, so that $T = \frac{60''}{n}$, then

$$P = \frac{39{\cdot}4784}{3600} \times n^2 \times M r = {\cdot}010966\, n^2 \times M r,$$

$$\text{or } P = {\cdot}000331 \times n^2 \times G r.$$

Lastly, as $\frac{2\pi}{T} = \omega$, the *angular* velocity,

$$P = \omega_2 \times M r = \omega^2 \times \frac{G}{g} r.$$

These various expressions for P become convenient for all calculations in which centrifugal force enters.

Where the revolving body is a fluid, and a particle whose weight is G is transferred by the normal force from the axis of rotation to the extremity of the radius r, then the work done, L, is—

$$L = \frac{\omega^2 r^2}{2g} \times G = \frac{v^2}{2g} G,$$

v being the velocity of rotation at the *extremity* of the radius r. Or if the particle start not from the axis but from some intermediate point in a radius, then

$$L = \left(\frac{v_2 - v_{12}}{2g}\right) G$$

In the case of a properly proportioned centrifugal pump—

Let $CR = r_1$.

v_1 = the surface velocity of the vanes, which must be proportioned to

H = maximum dynamic head of water to be overcome, and which consists of

z = the elevation to which the water is to be delivered from the lower level.

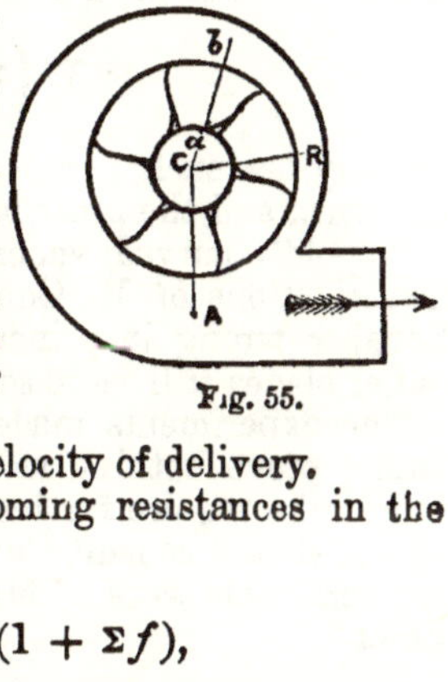

Fig. 55.

h = the height due to the velocity of delivery.

h_1 = the head lost in overcoming resistances in the machine.

$$\text{Then } \frac{v_{12}}{g} = H = z + \frac{V^2}{2g}(1 + \Sigma f),$$

V being the velocity in the ascending main of the pump, and Σf the sum of the several resistances; and the surface velocity of the blades is—

$$v_1 = \sqrt{\left(gz + \frac{V^2}{2}\left(1 + \cdot 025 \frac{z}{d}\right)\right)}$$

d being the diameter of the pipe when it is wholly vertical, and therefore its length $l = z$; but when otherwise for $\cdot 025 \frac{z}{d}$ we must substitute $\cdot 025 \frac{l}{d}$

Let d, = the diameter of the ascending main, be taken as unity. Then in proportioning the pump, let the external radius of the blades, CR (Fig. 55), $= \frac{7}{4} d$; the radius of the ears of the pump $= \frac{3}{4} d$; and the diameter of each of the indraught passages $= d$. The breadth of the blades $= \frac{3}{4} d$ nearly, and the mean radius of the casing of the pump $= \frac{7}{4} d \times \frac{v_1}{V} =$ CA, Fig. 1.

The fan-blades should be perfectly radial at the outer extremity, and for at least one-half their length. The inner portion should be curved, as in Fig. 56, forwards in the direction of revolution of the fan, and should so reach the inner edge of the revolving disc of blades, that the angle $p\,o\,s$, $= \beta$, should be—

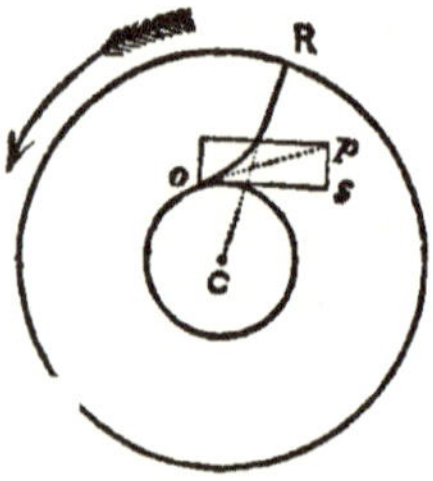

Fig. 56.

$$\beta = \frac{V_0}{V_2}$$

V_0 being the radial velocity of the water, and V_2 being that of the inner edge of the fan-blade.

As regards the power required to drive a centrifugal pump, and to raise per minute a given weight of water, W, it may be taken at

$$2\,W\left(z + \frac{V^2}{2g}\left(1 + \cdot 025\frac{l}{d}\right)\right)$$

for very few such pumps in reality return in duty more than fifty per cent., and the great majority far less.

Appold's curved vanes, though apparently sanctioned by the investigations of M. Combes on fans for ventilation, are unquestionably wrong in principle and in practice, and the uncurved radial blades still more so in both.

The experiments made in 1851, and referred to in the text, in reality prove nothing as to the general superiority of the former. Those who require further information on the subject of centrifugal pumps should consult Professor Rankine's Applied Mechanics, and the admirable work of Morin, "Des Machines pour l'élévation des Eaux."

INDEX.

THE END.

PRINTED BY J. S. VIRTUE AND CO., LIMITED, CITY ROAD, LONDON.

Weale's Rudimentary Series.

LONDON, 1862.
THE PRIZE MEDAL
Was awarded to the Publishers of
"WEALE'S SERIES."

A NEW LIST OF

WEALE'S SERIES

RUDIMENTARY SCIENTIFIC, EDUCATIONAL, AND CLASSICAL.

Comprising nearly Three Hundred and Fifty distinct works in almost every department of Science, Art, and Education, recommended to the notice of Engineers, Architects, Builders, Artisans, and Students generally, as well as to those interested in Workmen's Libraries, Literary and Scientific Institutions, Colleges, Schools, Science Classes, &c., &c.

☞ "WEALE'S SERIES includes Text-Books on almost every branch of Science and Industry, comprising such subjects as Agriculture, Architecture and Building, Civil Engineering, Fine Arts, Mechanics and Mechanical Engineering, Physical and Chemical Science, and many miscellaneous Treatises. The whole are constantly undergoing revision, and new editions, brought up to the latest discoveries in scientific research, are constantly issued. The prices at which they are sold are as low as their excellence is assured."—*American Literary Gazette.*

"Amongst the literature of technical education, WEALE'S SERIES has ever enjoyed a high reputation, and the additions being made by Messrs. CROSBY LOCKWOOD & CO. render the series even more complete, and bring the information upon the several subjects down to the present time."—*Mining Journal.*

"It is not too much to say that no books have ever proved more popular with, or more useful to, young engineers and others than the excellent treatises comprised in WEALE'S SERIES."—*Engineer.*

"The excellence of WEALE'S SERIES is now so well appreciated, that it would be wasting our space to enlarge upon their general usefulness and value."—*Builder.*

"WEALE'S SERIES has become a standard as well as an unrivalled collection of treatises in all branches of art and science."—*Public Opinion.*

PHILADELPHIA, 1876.
THE PRIZE MEDAL
Was awarded to the Publishers for
Books: Rudimentary, Scientific,
"WEALE'S SERIES," ETC.

CROSBY LOCKWOOD & CO.,

7, STATIONERS' HALL COURT, LUDGATE HILL, LONDON, E.C.

WEALE'S RUDIMENTARY SCIENTIFIC SERIES.

*** The Volumes of this Series are freely Illustrated with Woodcuts, or otherwise, where requisite. Throughout the following List it must be understood that the books are bound in limp cloth, unless otherwise stated; *but the volumes marked with a ‡ may also be had strongly bound in cloth boards for 6d. extra.*

N.B.—In ordering from this List it is recommended, as a means of facilitating business and obviating error, to quote the numbers affixed to the volumes, as well as the titles and prices.

CIVIL ENGINEERING, SURVEYING, ETC.

No.

31. *WELLS AND WELL-SINKING.* By JOHN GEO. SWINDELL, A.R.I.B.A., and G. R. BURNELL, C.E. Revised Edition. With a New Appendix on the Qualities of Water. Illustrated. 2s.

35. *THE BLASTING AND QUARRYING OF STONE,* for Building and other Purposes. With Remarks on the Blowing up of Bridges. By Gen. Sir JOHN BURGOYNE, Bart., K.C.B. Illustrated. 1s. 6d.

43. *TUBULAR, AND OTHER IRON GIRDER BRIDGES,* particularly describing the Britannia and Conway Tubular Bridges. By G. DRYSDALE DEMPSEY, C.E. Fourth Edition. 2s.

44. *FOUNDATIONS AND CONCRETE WORKS,* with Practical Remarks on Footings, Sand, Concrete, Béton, Pile-driving, Caissons, and Cofferdams, &c. By E. DOBSON. Fifth Edition. 1s. 6d.

60. *LAND AND ENGINEERING SURVEYING.* By T. BAKER, C.E. New Edition, revised by EDWARD NUGENT, C.E. 2s.‡

80*. *EMBANKING LANDS FROM THE SEA.* With examples and Particulars of actual Embankments, &c. By J. WIGGINS, F.G.S. 2s.

81. *WATER WORKS,* for the Supply of Cities and Towns. With a Description of the Principal Geological Formations of England as influencing Supplies of Water; and Details of Engines and Pumping Machinery for raising Water. By SAMUEL HUGHES, F.G.S., C.E. New Edition. 4s.‡

118. *CIVIL ENGINEERING IN NORTH AMERICA,* a Sketch of. By DAVID STEVENSON, F.R.S.E., &c. Plates and Diagrams. 3s.

167. *IRON BRIDGES, GIRDERS, ROOFS, AND OTHER WORKS.* By FRANCIS CAMPIN, C.E. 2s. 6d.‡

197. *ROADS AND STREETS (THE CONSTRUCTION OF).* By HENRY LAW, C.E., revised and enlarged by D. K. CLARK, C.E., including pavements of Stone, Wood, Asphalte, &c. 4s. 6d.‡

203. *SANITARY WORK IN THE SMALLER TOWNS AND IN VILLAGES.* By C. SLAGG, A.M.I.C.E. Revised Edition. 3s.‡

212. *GAS-WORKS, THEIR CONSTRUCTION AND ARRANGEMENT;* and the Manufacture and Distribution of Coal Gas. Originally written by SAMUEL HUGHES, C.E. Re-written and enlarged by WILLIAM RICHARDS, C.E. Seventh Edition, with important additions. 5s. 6d.‡

213. *PIONEER ENGINEERING.* A Treatise on the Engineering Operations connected with the Settlement of Waste Lands in New Countries. By EDWARD DOBSON, Assoc. Inst. C.E. 4s. 6d.‡

216. *MATERIALS AND CONSTRUCTION;* A Theoretical and Practical Treatise on the Strains, Designing, and Erection of Works of Construction. By FRANCIS CAMPIN, C.E. Second Edition, revised. 3s.‡

219. *CIVIL ENGINEERING.* By HENRY LAW, M.Inst. C.E. Including HYDRAULIC ENGINEERING by GEO. R. BURNELL, M.Inst. C.E. Seventh Edition, revised, with large additions by D. KINNEAR CLARK, M.Inst. C.E. 6s. 6d., Cloth boards, 7s. 6d.

☞ *The ‡ indicates that these vols. may be had strongly bound at 6d. extra.*

MECHANICAL ENGINEERING, ETC.

33. *CRANES*, the Construction of, and other Machinery for Raising Heavy Bodies. By JOSEPH GLYNN, F.R.S. Illustrated. 1s. 6d.

34. *THE STEAM ENGINE.* By Dr. LARDNER. Illustrated. 1s. 6d.

59. *STEAM BOILERS:* their Construction and Management. By R. ARMSTRONG, C.E. Illustrated. 1s. 6d.

82. *THE POWER OF WATER*, as applied to drive Flour Mills, and to give motion to Turbines, &c. By JOSEPH GLYNN, F.R.S. 2s.‡

98. *PRACTICAL MECHANISM*, the Elements of; and Machine Tools. By T. BAKER, C.E. With Additions by J. NASMYTH, C.E. 2s. 6d.‡

139. *THE STEAM ENGINE*, a Treatise on the Mathematical Theory of, with Rules and Examples for Practical Men. By T. BAKER, C.E. 1s. 6d.

164. *MODERN WORKSHOP PRACTICE*, as applied to Steam Engines, Bridges, Cranes, Ship-building, &c. By J. G. WINTON. 3s.‡

165. *IRON AND HEAT*, exhibiting the Principles concerned in the Construction of Iron Beams, Pillars, and Girders. By J. ARMOUR. 2s. 6d.‡

166. *POWER IN MOTION:* Horse-Power, Toothed-Wheel Gearing, Long and Short Driving Bands, and Angular Forces. By J. ARMOUR, 2s. 6d.‡

171. *THE WORKMAN'S MANUAL OF ENGINEERING* DRAWING. By J. MAXTON. 6th Edn. With 7 Plates and 350 Cuts. 3s. 6d.‡

190. *STEAM AND THE STEAM ENGINE*, Stationary and Portable. By JOHN SEWELL and D. K. CLARK, M.I.C.E. 3s. 6d.‡

200. *FUEL*, its Combustion and Economy. By C. W. WILLIAMS, With Recent Practice in the Combustion and Economy of Fuel—Coal, Coke, Wood, Peat, Petroleum, &c.—by D. K. CLARK, M.I.C.E. 3s. 6d.‡

202. *LOCOMOTIVE ENGINES.* By G. D. DEMPSEY, C.E.; with large additions by D. KINNEAR CLARK, M.I.C.E. 3s.‡

211. *THE BOILERMAKER'S ASSISTANT* in Drawing, Templating, and Calculating Boiler and Tank Work. By JOHN COURTNEY. Practical Boiler Maker. Edited by D. K. CLARK, C.E. 100 Illustrations. 2s.

217. *SEWING MACHINERY:* Its Construction, History, &c., with full Technical Directions for Adjusting, &c. By J. W. URQUHART, C.E. 2s.‡

223. *MECHANICAL ENGINEERING.* Comprising Metallurgy, Moulding, Casting, Forging, Tools, Workshop Machinery, Manufacture of the Steam Engine, &c. By FRANCIS CAMPIN, C.E. 2s. 6d.‡

236. *DETAILS OF MACHINERY.* Comprising Instructions for the Execution of various Works in Iron. By FRANCIS CAMPIN, C.E. 3s.‡

237. *THE SMITHY AND FORGE;* including the Farrier's Art and Coach Smithing. By W. J. E. CRANE. Illustrated. 2s. 6d.‡

238. *THE SHEET-METAL WORKER'S GUIDE;* a Practical Handbook for Tinsmiths, Coppersmiths, Zincworkers, &c. With 94 Diagrams and Working Patterns. By W. J. E. CRANE. 1s. 6d.

251. *STEAM AND MACHINERY MANAGEMENT:* with Hints on Construction and Selection. By M. POWIS BALE, M.I.M.E. 2s. 6d.‡

254. *THE BOILERMAKER'S READY-RECKONER.* By J. COURTNEY. Edited by D. K. CLARK, C.E. 4s., limp; 5s., half-bound.

255. *LOCOMOTIVE ENGINE-DRIVING.* A Practical Manual for Engineers in charge of Locomotive Engines. By MICHAEL REYNOLDS, M.S.E. Seventh Edition. 3s. 6d., limp; 4s. 6d. cloth boards.

256. *STATIONARY ENGINE-DRIVING.* A Practical Manual for Engineers in charge of Stationary Engines. By MICHAEL REYNOLDS, M.S.E. Third Edition. 3s. 6d. limp; 4s. 6d. cloth boards.

260. *IRON BRIDGES OF MODERATE SPAN:* their Construction and Erection. By HAMILTON W. PENDRED, late Inspector of Ironwork to the Salford Corporation. 2s. [*Just published.*

☞ *The ‡ indicates that these vols. may be had strongly bound at 6d. extra.*

MINING, METALLURGY, ETC.

4. *MINERALOGY*, Rudiments of; a concise View of the General Properties of Minerals. By A. RAMSAY, F.G.S., F.R.G.S., &c. Third Edition, revised and enlarged. Illustrated. 3s. 6d.‡

117. *SUBTERRANEOUS SURVEYING*, with and without the Magnetic Needle. By T. FENWICK and T. BAKER, C.E. Illustrated. 2s. 6d.‡

133. *METALLURGY OF COPPER*; an Introduction to the Methods of Seeking, Mining, and Assaying Copper, and Manufacturing its Alloys. By ROBERT H. LAMBORN, Ph.D. Woodcuts. 2s. 6d.‡

135. *ELECTRO-METALLURGY*; Practically Treated. By ALEXANDER WATT, F.R.S.S.A. Eighth Edition, revised, with additional Matter and Illustrations, including the most recent Processes. 3s.‡

172. *MINING TOOLS*, Manual of. For the Use of Mine Managers, Agents, Students, &c. By WILLIAM MORGANS. 2s. 6d.‡

172*. *MINING TOOLS, ATLAS* of Engravings to Illustrate the above, containing 235 Illustrations, drawn to Scale. 4to. 4s. 6d.; cloth boards, 6s.

176. *METALLURGY OF IRON*. Containing History of Iron Manufacture, Methods of Assay, and Analyses of Iron Ores, Processes of Manufacture of Iron and Steel, &c. By H. BAUERMAN, F.G.S. Fifth Edition, revised and enlarged. 5s.‡

180. *COAL AND COAL MINING*. By WARINGTON W. SMYTH, M.A., F.R.S. Sixth Edition, revised. 3s. 6d.‡

195. *THE MINERAL SURVEYOR AND VALUER'S COMPLETE GUIDE*, with new Traverse Tables, and Descriptions of Improved Instruments; also the Correct Principles of Laying out and Valuing Mineral Properties. By WILLIAM LINTERN, Mining and Civil Engineer. 3s. 6d.‡

214. *SLATE AND SLATE QUARRYING*, Scientific, Practical, and Commercial. By D. C. DAVIES, F.G.S., Mining Engineer, &c. 3s.‡

220. *MAGNETIC SURVEYING, AND ANGULAR SURVEYING*, with Records of the Peculiarities of Needle Disturbances. Compiled from the Results of carefully made Experiments. By W. LINTERN. 2s.

ARCHITECTURE, BUILDING, ETC.

16. *ARCHITECTURE—ORDERS*—The Orders and their Æsthetic Principles. By W. H. LEEDS. Illustrated. 1s. 6d.

17. *ARCHITECTURE—STYLES*—The History and Description of the Styles of Architecture of Various Countries, from the Earliest to the Present Period. By T. TALBOT BURY, F.R.I.B.A., &c. Illustrated. 2s.
*** ORDERS AND STYLES OF ARCHITECTURE, *in One Vol.*, 3s. 6d.

18. *ARCHITECTURE—DESIGN*—The Principles of Design in Architecture, as deducible from Nature and exemplified in the Works of the Greek and Gothic Architects. By E. L. GARBETT, Architect. Illustrated. 2s. 6d.

*** *The three preceding Works, in One handsome Vol., half bound, entitled* "MODERN ARCHITECTURE," *price* 6s.

22. *THE ART OF BUILDING*, Rudiments of. General Principles of Construction, Materials used in Building, Strength and Use of Materials, Working Drawings, Specifications, and Estimates. By E. DOBSON, 2s.‡

25. *MASONRY AND STONECUTTING*: Rudimentary Treatise on the Principles of Masonic Projection and their application to Construction. By EDWARD DOBSON, M.R.I.B.A., &c. 2s. 6d.‡

42. *COTTAGE BUILDING*. By C. BRUCE ALLEN, Architect. Tenth Edition, revised and enlarged. With a Chapter on Economic Cottages for Allotments, by EDWARD E. ALLEN, C.E. 2s.

45. *LIMES, CEMENTS, MORTARS, CONCRETES, MASTICS, PLASTERING*, &c. By G. R. BURNELL, C.E. Twelfth Edition. 1s. 6d.

☞ *The ‡ indicates that these vols. may be had strongly bound at 6d. extra.*

Architecture, Building, etc., *continued.*

57. *WARMING AND VENTILATION.* An Exposition of the General Principles as applied to Domestic and Public Buildings, Mines, Lighthouses, Ships, &c. By C. TOMLINSON, F.R.S., &c. Illustrated. 3s.

111. *ARCHES, PIERS, BUTTRESSES, &c.:* Experimental Essays on the Principles of Construction. By W. BLAND. Illustrated. 1s. 6d.

116. *THE ACOUSTICS OF PUBLIC BUILDINGS;* or, The Principles of the Science of Sound applied to the purposes of the Architect and Builder. By T. ROGER SMITH, M.R.I.B.A., Architect. Illustrated. 1s. 6d.

127. *ARCHITECTURAL MODELLING IN PAPER,* the Art of. By T. A. RICHARDSON, Architect. Illustrated. 1s. 6d.

128. *VITRUVIUS—THE ARCHITECTURE OF MARCUS VITRUVIUS POLLO.* In Ten Books. Translated from the Latin by JOSEPH GWILT, F.S.A., F.R.A.S. With 23 Plates. 5s.

130. *GRECIAN ARCHITECTURE,* An Inquiry into the Principles of Beauty in; with an Historical View of the Rise and Progress of the Art in Greece. By the EARL OF ABERDEEN. 1s.

⁎⁎ The two preceding Works in One handsome Vol., half bound, entitled "ANCIENT ARCHITECTURE," *price* 6*s.*

132. *THE ERECTION OF DWELLING-HOUSES.* Illustrated by a Perspective View, Plans, Elevations, and Sections of a pair of Semi-detached Villas, with the Specification, Quantities, and Estimates, &c. By S. H. BROOKS. New Edition, with Plates. 2s. 6d.‡

156. *QUANTITIES & MEASUREMENTS* in Bricklayers', Masons', Plasterers', Plumbers', Painters', Paperhangers', Gilders', Smiths', Carpenters' and Joiners' Work. By A. C. BEATON, Surveyor. New Edition. 1s. 6d.

175. *LOCKWOOD & CO.'S BUILDER'S AND CONTRACTOR'S* PRICE BOOK, containing the latest Prices of all kinds of Builders' Materials and Labour, and of all Trades connected with Building, &c., &c. Edited by F. T. W. MILLER, Architect. Published annually. 3s. 6d.; half bound, 4s.

182. *CARPENTRY AND JOINERY*—THE ELEMENTARY PRINCIBLES OF CARPENTRY. Chiefly composed from the Standard Work of THOMAS TREDGOLD, C.E. With a TREATISE ON JOINERY by E. WYNDHAM TARN, M.A. Fourth Edition, Revised. 3s. 6d.‡

182*. *CARPENTRY AND JOINERY. ATLAS* of 35 Plates to accompany the above. With Descriptive Letterpress. 4to. 6s.

185. *THE COMPLETE MEASURER;* the Measurement of Boards, Glass, &c.; Unequal-sided, Square-sided, Octagonal-sided, Round Timber and Stone, and Standing Timber, &c. By RICHARD HORTON. Fifth Edition. 4s.; strongly bound in leather, 5s.

187. *HINTS TO YOUNG ARCHITECTS.* By G. WIGHTWICK. New Edition. By G. H. GUILLAUME. Illustrated. 3s. 6d.‡

188. *HOUSE PAINTING, GRAINING, MARBLING, AND SIGN WRITING:* with a Course of Elementary Drawing for House-Painters, Sign-Writers, &c., and a Collection of Useful Receipts. By ELLIS A. DAVIDSON. Fourth Edition. With Coloured Plates. 5s. cloth limp; 6s. cloth boards.

189. *THE RUDIMENTS OF PRACTICAL BRICKLAYING.* In Six Sections: General Principles; Arch Drawing, Cutting, and Setting; Pointing; Paving, Tiling, Materials; Slating and Plastering; Practical Geometry, Mensuration, &c. By ADAM HAMMOND. Sixth Edition. 1s. 6d.

191. *PLUMBING.* A Text-Book to the Practice of the Art or Craft of the Plumber. With Chapters upon House Drainage. Fourth Edition. With 330 Illustrations. By W. P. BUCHAN. 3s. 6d.‡

192. *THE TIMBER IMPORTER'S, TIMBER MERCHANT'S,* and BUILDER'S STANDARD GUIDE. By R. E. GRANDY. 3s.‡

206. *A BOOK ON BUILDING, Civil and Ecclesiastical,* including CHURCH RESTORATION. With the Theory of Domes and the Great Pyramid, &c. By Sir EDMUND BECKETT, Bart., LL.D., Q.C., F.R.A.S. 4s. 6d.‡

☞ *The ‡ indicates that these vols. may be had strongly bound at 6d. extra.*

Architecture, Building, etc., *continued.*

226. *THE JOINTS MADE AND USED BY BUILDERS* in the Construction of various kinds of Engineering and Architectural Works. By WYVILL J. CHRISTY, Architect. With upwards of 160 Engravings on Wood. 3s.‡

228. *THE CONSTRUCTION OF ROOFS OF WOOD AND IRON* By E. WYNDHAM TARN, M.A., Architect. Second Edition, revised. 1s. 6d.

229. *ELEMENTARY DECORATION:* as applied to the Interior and Exterior Decoration of Dwelling-Houses, &c. By J. W. FACEY. 2s.

257. *PRACTICAL HOUSE DECORATION.* A Guide to the Art of Ornamental Painting, the Arrangement of Colours in Apartments, and the Principles of Decorative Design. With Remarks on the Nature and Properties of Pigments. By JAMES W. FACEY. 2s. 6d.

*** *The two preceding Works, in One handsome Vol., half-bound, entitled* "HOUSE DECORATION, ELEMENTARY AND PRACTICAL," *price* 5s.

230. *HANDRAILING.* Showing New and Simple Methods for finding the Pitch of the Plank, Drawing the Moulds, Bevelling, Jointing-up, and Squaring the Wreath. By GEORGE COLLINGS. Plates and Diagrams. 1s. 6d.

247. *BUILDING ESTATES:* a Rudimentary Treatise on the Development, Sale, Purchase, and General Management of Building Land. By FOWLER MAITLAND, Surveyor. Illustrated. 2s.

248. *PORTLAND CEMENT FOR USERS.* By HENRY FAIJA, Assoc. M. Inst. C.E. Second Edition, corrected. Illustrated. 2s.

252. *BRICKWORK:* a Practical Treatise, embodying the General and Higher Principles of Bricklaying, Cutting and Setting, &c. By F. WALKER. Second Edition, Revised and Enlarged. 1s. 6d.

23. *THE PRACTICAL BRICK AND TILE BOOK.* Comprising:
189. BRICK AND TILE MAKING, by E. DOBSON, A.I.C.E.; PRACTICAL BRICKLAY-
252. ING, by A. HAMMOND; BRICKWORK, by F. WALKER. 550 pp. with 270 Illustrations. 6s. Strongly half-bound.

253. *THE TIMBER MERCHANT'S, SAW-MILLER'S, AND* IMPORTER'S FREIGHT-BOOK AND ASSISTANT. By WM. RICHARDSON. With a Chapter on Speeds of Saw-Mill Machinery, &c. By M. POWIS BALE, A.M.Inst.C.E. 3s.‡

250. *CIRCULAR WORK IN CARPENTRY AND JOINERY.* A Practical Treatise on Circular Work of Single and Double Curvature. By GEORGE COLLINGS, Author of "A Practical Treatise on Handrailing." 2s. 6d. [*Just published.*

259. *GAS FITTING:* A Practical Handbook treating of every Description of Gas Laying and Fitting. By JOHN BLACK. With 122 Illustrations. 2s. 6d.‡ [*Just published.*

261. *SHORING AND ITS APPLICATION:* A Handbook for the Use of Students. By GEORGE H. BLAGROVE. 1s. 6d. [*Just published.*

SHIPBUILDING, NAVIGATION, MARINE ENGINEERING, ETC.

51. *NAVAL ARCHITECTURE.* An Exposition of the Elementary Principles of the Science, and their Practical Application to Naval Construction. By J. PEAKE. Fifth Edition, with Plates and Diagrams. 3s. 6d.‡

53*. *SHIPS FOR OCEAN & RIVER SERVICE,* Elementary and Practical Principles of the Construction of. By H. A. SOMMERFELDT. 1s. 6d.

53**. *AN ATLAS OF ENGRAVINGS* to Illustrate the above. Twelve large folding plates. Royal 4to, cloth. 7s. 6d.

54. *MASTING, MAST-MAKING, AND RIGGING OF SHIPS,* Also Tables of Spars, Rigging, Blocks; Chain, Wire, and Hemp Ropes, &c., relative to every class of vessels. By ROBERT KIPPING, N.A. 2s.‡

54*. *IRON SHIP-BUILDING.* With Practical Examples and Details. By JOHN GRANTHAM, C.E. 5th Edition. 4s.

☞ *The ‡ indicates that these vols. may be had strongly bound at 6d. extra.*

Shipbuilding, Navigation, Marine Engineering, etc., *cont.*

55. *THE SAILOR'S SEA BOOK:* a Rudimentary Treatise on Navigation. By JAMES GREENWOOD, B.A. With numerous Woodcuts and Coloured Plates. New and enlarged edition. By W. H. ROSSER. 2s. 6d.‡

80. *MARINE ENGINES AND STEAM VESSELS.* By ROBERT MURRAY, C.E. Eighth Edition, thoroughly Revised, with Additions by the Author and by GEORGE CARLISLE, C.E., Senior Surveyor to the Board of Trade, Liverpool. 4s. 6d. limp; 5s. cloth boards. [*Just published.*

83*bis.* *THE FORMS OF SHIPS AND BOATS.* By W. BLAND. Seventh Edition, Revised, with numerous Illustrations and Models. 1s. 6d.

99. *NAVIGATION AND NAUTICAL ASTRONOMY,* in Theory and Practice. By Prof. J. R. YOUNG. New Edition. 2s. 6d.

106. *SHIPS' ANCHORS,* a Treatise on. By G. COTSELL, N.A. 1s. 6d.

149. *SAILS AND SAIL-MAKING.* With Draughting, and the Centre of Effort of the Sails; Weights and Sizes of Ropes; Masting, Rigging, and Sails of Steam Vessels, &c. 11th Edition. By R. KIPPING, N.A., 2s. 6d.‡

155. *ENGINEER'S GUIDE TO THE ROYAL & MERCANTILE* NAVIES. By a PRACTICAL ENGINEER. Revised by D. F. M'CARTHY. 3s.

55 & 204. *PRACTICAL NAVIGATION.* Consisting of The Sailor's Sea-Book. By JAMES GREENWOOD and W. H. ROSSER. Together with the requisite Mathematical and Nautical Tables for the Working of the Problems. By H. LAW, C.E., and Prof. J. R. YOUNG. 7s. Half-bound.

AGRICULTURE, GARDENING, ETC.

61*. *A COMPLETE READY RECKONER FOR THE ADMEA-*SUREMENT OF LAND, &c. By A. ARMAN. Second Edition, revised and extended by C. NORRIS, Surveyor, Valuer, &c. 2s.

131. *MILLER'S, CORN MERCHANT'S, AND FARMER'S* READY RECKONER. Second Edition, with a Price List of Modern Flour-Mill Machinery, by W. S. HUTTON, C.E. 2s.

140. *SOILS, MANURES, AND CROPS.* (Vol. 1. OUTLINES OF MODERN FARMING.) By R. SCOTT BURN. Woodcuts. 2s.

141. *FARMING & FARMING ECONOMY,* Notes, Historical and Practical, on. (Vol. 2. OUTLINES OF MODERN FARMING.) By R. SCOTT BURN. 3s.

142. *STOCK; CATTLE, SHEEP, AND HORSES.* (Vol. 3. OUTLINES OF MODERN FARMING.) By R. SCOTT BURN. Woodcuts. 2s. 6d.

145. *DAIRY, PIGS, AND POULTRY,* Management of the. By R. SCOTT BURN. (Vol. 4. OUTLINES OF MODERN FARMING.) 2s.

146. *UTILIZATION OF SEWAGE, IRRIGATION, AND* RECLAMATION OF WASTE LAND. (Vol. 5. OUTLINES OF MODERN FARMING.) By R. SCOTT BURN. Woodcuts. 2s. 6d.

*** *Nos.* 140-1-2-5-6, *in One Vol., handsomely half-bound, entitled* "OUTLINES OF MODERN FARMING." By ROBERT SCOTT BURN. *Price* 12*s.*

177. *FRUIT TREES,* The Scientific and Profitable Culture of. From the French of DU BREUIL. Revised by GEO. GLENNY. 187 Woodcuts. 3s. 6d.‡

198. *SHEEP; THE HISTORY, STRUCTURE, ECONOMY, AND* DISEASES OF. By W. C. SPOONER, M.R.V.C., &c. Fourth Edition, enlarged, including Specimens of New and Improved Breeds. 3s. 6d.‡

201. *KITCHEN GARDENING MADE EASY.* By GEORGE M. F. GLENNY. Illustrated. 1s. 6d.‡

207. *OUTLINES OF FARM MANAGEMENT, and the Organization of Farm Labour.* By R. SCOTT BURN. 2s. 6d.‡

208. *OUTLINES OF LANDED ESTATES MANAGEMENT.* By R. SCOTT BURN. 2s. 6d.‡

*** *Nos.* 207 & 208 *in One Vol., handsomely half bound, entitled* "OUTLINES OF LANDED ESTATES AND FARM MANAGEMENT." By R. SCOTT BURN. *Price* 6*s.*

☞ *The ‡ indicates that these vols. may be had strongly bound at 6d. extra.*

Agriculture, Gardening, etc., *continued.*

209. *THE TREE PLANTER AND PLANT PROPAGATOR.* A Practical Manual on the Propagation of Forest Trees, Fruit Trees, Flowering Shrubs, Flowering Plants, &c. By SAMUEL WOOD. 2s.‡

210. *THE TREE PRUNER.* A Practical Manual on the Pruning of Fruit Trees, including also their Training and Renovation; also the Pruning of Shrubs, Climbers, and Flowering Plants. By SAMUEL WOOD. 2s.‡

⁂ *Nos.* 209 *&* 210 *in One Vol., handsomely half-bound, entitled* "THE TREE PLANTER, PROPAGATOR, AND PRUNER." By SAMUEL WOOD. *Price* 5*s.*

218. *THE HAY AND STRAW MEASURER:* Being New Tables for the Use of Auctioneers, Valuers, Farmers, Hay and Straw Dealers, &c. By JOHN STEELE. Fourth Edition. 2s.

222. *SUBURBAN FARMING.* The Laying-out and Cultivation of Farms, adapted to the Produce of Milk, Butter, and Cheese, Eggs, Poultry, and Pigs. By Prof. JOHN DONALDSON and R. SCOTT BURN. 3s. 6d.‡

231. *THE ART OF GRAFTING AND BUDDING.* By CHARLES BALTET. With Illustrations. 2s. 6d.‡

232. *COTTAGE GARDENING;* or, Flowers, Fruits, and Vegetables for Small Gardens. By E. HOBDAY. 1s. 6d.

233. *GARDEN RECEIPTS.* Edited by CHARLES W. QUIN. 1s. 6d.

234. *THE KITCHEN AND MARKET GARDEN.* Compiled by C. W. SHAW, Editor of "Gardening Illustrated." 3s.‡

239. *DRAINING AND EMBANKING.* A Practical Treatise, embodying the most recent experience in the Application of Improved Methods. By JOHN SCOTT, late Professor of Agriculture and Rural Economy at the Royal Agricultural College, Cirencester. With 68 Illustrations. 1s. 6d.

240. *IRRIGATION AND WATER SUPPLY.* A Treatise on Water Meadows, Sewage Irrigation, Warping, &c.; on the Construction of Wells, Ponds, and Reservoirs; and on Raising Water by Machinery for Agricultural and Domestic Purposes. By Prof. JOHN SCOTT. With 34 Illus. 1s. 6d.

241. *FARM ROADS, FENCES, AND GATES.* A Practical Treatise on the Roads, Tramways, and Waterways of the Farm; the Principles of Enclosures; and the different kinds of Fences, Gates, and Stiles. By Professor JOHN SCOTT. With 75 Illustrations. 1s. 6d.

242. *FARM BUILDINGS.* A Practical Treatise on the Buildings necessary for various kinds of Farms, their Arrangement and Construction, including Plans and Estimates. By Prof. JOHN SCOTT. With 105 Illus. 2s.

243. *BARN IMPLEMENTS AND MACHINES.* A Practical Treatise on the Application of Power to the Operations of Agriculture; and on various Machines used in the Threshing-barn, in the Stock-yard, and in the Dairy, &c. By Prof. J. SCOTT. With 123 Illustrations. 2s.

244. *FIELD IMPLEMENTS AND MACHINES.* A Practical Treatise on the Varieties now in use, with Principles and Details of Construction, their Points of Excellence, and Management. By Professor JOHN SCOTT. With 138 Illustrations. 2s.

245. *AGRICULTURAL SURVEYING.* A Practical Treatise on Land Surveying, Levelling, and Setting-out; and on Measuring and Estimating Quantities, Weights, and Values of Materials, Produce, Stock, &c. By Prof. JOHN SCOTT. With 62 Illustrations. 1s. 6d.

⁂ *Nos.* 239 *to* 245 *in One Vol., handsomely half-bound, entitled* "THE COMPLETE TEXT-BOOK OF FARM ENGINEERING." By Professor JOHN SCOTT. *Price* 12*s.*

250. *MEAT PRODUCTION.* A Manual for Producers, Distributors, &c. By JOHN EWART. 2s. 6d.‡

☞ *The ‡ indicates that these vols. may be had strongly bound at 6d. extra.*

MATHEMATICS, ARITHMETIC, ETC.

32. *MATHEMATICAL INSTRUMENTS,* a Treatise on; in which their Construction and the Methods of Testing, Adjusting, and Using them are concisely Explained. By J. F. HEATHER, M.A., of the Royal Military Academy, Woolwich. Original Edition, in 1 vol., Illustrated. 1s. 6d.

**** *In ordering the above, be careful to say, " Original Edition " (No. 32), to distinguish it from the Enlarged Edition in 3 vols. (Nos. 168-9-70.)*

76. *DESCRIPTIVE GEOMETRY,* an Elementary Treatise on; with a Theory of Shadows and of Perspective, extracted from the French of G. MONGE. To which is added, a description of the Principles and Practice of Isometrical Projection. By J. F. HEATHER, M.A. With 14 Plates. 2s.

178. *PRACTICAL PLANE GEOMETRY:* giving the Simplest Modes of Constructing Figures contained in one Plane and Geometrical Construction of the Ground. By J. F. HEATHER, M.A. With 215 Woodcuts. 2s.

83. *COMMERCIAL BOOK-KEEPING.* With Commercial Phrases and Forms in English, French, Italian, and German. By JAMES HADDON, M.A., Arithmetical Master of King's College School, London. 1s. 6d.

84. *ARITHMETIC,* a Rudimentary Treatise on: with full Explanations of its Theoretical Principles, and numerous Examples for Practice. By Professor J. R. YOUNG. Tenth Edition, corrected. 1s. 6d.

84*. A KEY to the above, containing Solutions in full to the Exercises, together with Comments, Explanations, and Improved Processes, for the Use of Teachers and Unassisted Learners. By J. R. YOUNG. 1s. 6d.

85. *EQUATIONAL ARITHMETIC,* applied to Questions of Interest, Annuities, Life Assurance, and General Commerce; with various Tables by which all Calculations may be greatly facilitated. By W. HIPSLEY. 2s.

86. *ALGEBRA,* the Elements of. By JAMES HADDON, M.A. With Appendix, containing miscellaneous Investigations, and a Collection of Problems in various parts of Algebra. 2s.

86*. A KEY AND COMPANION to the above Book, forming an extensive repository of Solved Examples and Problems in Illustration of the various Expedients necessary in Algebraical Operations. By J. R. YOUNG. 1s. 6d.

88. *EUCLID,* THE ELEMENTS OF: with many additional Propositions
89. and Explanatory Notes: to which is prefixed, an Introductory Essay on Logic. By HENRY LAW, C.E. 2s. 6d.‡

**** *Sold also separately, viz.:—*

88. EUCLID, The First Three Books. By HENRY LAW, C.E. 1s. 6d.

89. EUCLID, Books 4, 5, 6, 11, 12. By HENRY LAW, C.E. 1s. 6d.

90. *ANALYTICAL GEOMETRY AND CONIC SECTIONS,* By JAMES HANN. A New Edition, by Professor J. R. YOUNG. 2s.‡

91. *PLANE TRIGONOMETRY,* the Elements of. By JAMES HANN, formerly Mathematical Master of King's College, London. 1s. 6d.

92. *SPHERICAL TRIGONOMETRY,* the Elements of. By JAMES HANN. Revised by CHARLES H. DOWLING, C.E. 1s.

**** *Or with " The Elements of Plane Trigonometry," in One Volume, 2s.* 6d.

93. *MENSURATION AND MEASURING.* With the Mensuration and Levelling of Land for the Purposes of Modern Engineering. By T. BAKER, C.E. New Edition by E. NUGENT, C.E. Illustrated. 1s. 6d.

101. *DIFFERENTIAL CALCULUS,* Elements of the. By W. S. B. WOOLHOUSE, F.R.A.S., &c. 1s. 6d.

102. *INTEGRAL CALCULUS,* Rudimentary Treatise on the. By HOMERSHAM COX, B.A. Illustrated. 1s.

105. *MNEMONICAL LESSONS.*—GEOMETRY, ALGEBRA, AND TRIGONOMETRY, in Easy Mnemonical Lessons. By the Rev. THOMAS PENYNGTON KIRKMAN, M.A. 1s. 6d.

136. *ARITHMETIC,* Rudimentary, for the Use of Schools and Self-Instruction. By JAMES HADDON, M.A. Revised by A. ARMAN. 1s. 6d.

137. A KEY TO HADDON'S RUDIMENTARY ARITHMETIC. By A. ARMAN. 1s. 6d.

☞ *The ‡ indicates that these vols. may be had strongly bound at 6d. extra.*

Mathematics, Arithmetic, etc., *continued.*

168. *DRAWING AND MEASURING INSTRUMENTS.* Including—I. Instruments employed in Geometrical and Mechanical Drawing, and in the Construction, Copying, and Measurement of Maps and Plans. II. Instruments used for the purposes of Accurate Measurement, and for Arithmetical Computations. By J. F. HEATHER, M.A. Illustrated. 1s. 6d.

169. *OPTICAL INSTRUMENTS.* Including (more especially) Telescopes, Microscopes, and Apparatus for producing copies of Maps and Plans by Photography. By J. F. HEATHER, M.A. Illustrated. 1s. 6d.

170. *SURVEYING AND ASTRONOMICAL INSTRUMENTS.* Including—I. Instruments Used for Determining the Geometrical Features of a portion of Ground. II. Instruments Employed in Astronomical Observations. By J. F. HEATHER, M.A. Illustrated. 1s. 6d.

*** The above three volumes form an enlargement of the Author's original work, "Mathematical Instruments." (See No. 32 in the Series.)*

168.
169.
170. *MATHEMATICAL INSTRUMENTS.* By J. F. HEATHER, M.A. Enlarged Edition, for the most part entirely re-written. The 3 Parts as above, in One thick Volume. With numerous Illustrations. 4s. 6d.‡

158. *THE SLIDE RULE, AND HOW TO USE IT;* containing full, easy, and simple Instructions to perform all Business Calculations with unexampled rapidity and accuracy. By CHARLES HOARE, C.E. Fifth Edition. With a Slide Rule in tuck of cover. 2s. 6d.‡

196. *THEORY OF COMPOUND INTEREST AND ANNUITIES;* with Tables of Logarithms for the more Difficult Computations of Interest, Discount, Annuities, &c. By FÉDOR THOMAN. 4s.‡

199. *THE COMPENDIOUS CALCULATOR;* or, Easy and Concise Methods of Performing the various Arithmetical Operations required in Commercial and Business Transactions; together with Useful Tables. By D. O'GORMAN. Twenty-sixth Edition, carefully revised by C. NORRIS. 3s., cloth limp; 3s. 6d., strongly half-bound in leather.

204. *MATHEMATICAL TABLES,* for Trigonometrical, Astronomical, and Nautical Calculations; to which is prefixed a Treatise on Logarithms. By HENRY LAW, C.E. Together with a Series of Tables for Navigation and Nautical Astronomy. By Prof. J. R. YOUNG. New Edition. 4s.‡

204*. *LOGARITHMS.* With Mathematical Tables for Trigonometrical, Astronomical, and Nautical Calculations. By HENRY LAW, M.Inst.C.E. New and Revised Edition. (Forming part of the above Work). 3s.

221. *MEASURES, WEIGHTS, AND MONEYS OF ALL NATIONS,* and an Analysis of the Christian, Hebrew, and Mahometan Calendars. By W. S. B. WOOLHOUSE, F.R.A.S., F.S.S. Sixth Edition. 2s.‡

227. *MATHEMATICS AS APPLIED TO THE CONSTRUCTIVE ARTS.* Illustrating the various processes of Mathematical Investigation, by means of Arithmetical and Simple Algebraical Equations and Practical Examples. By FRANCIS CAMPIN, C.E. Second Edition. 3s.‡

PHYSICAL SCIENCE, NATURAL PHILOSOPHY, ETC.

1. *CHEMISTRY.* By Professor GEORGE FOWNES, F.R.S. With an Appendix on the Application of Chemistry to Agriculture. 1s.

2. *NATURAL PHILOSOPHY,* Introduction to the Study of. By C. TOMLINSON. Woodcuts. 1s. 6d.

6. *MECHANICS,* Rudimentary Treatise on. By CHARLES TOMLINSON. Illustrated. 1s. 6d.

7. *ELECTRICITY;* showing the General Principles of Electrical Science, and the purposes to which it has been applied. By Sir W. SNOW HARRIS, F.R.S., &c. With Additions by R. SABINE, C.E., F.S.A. 1s. 6d.

7*. *GALVANISM.* By Sir W. SNOW HARRIS. New Edition by ROBERT SABINE, C.E., F.S.A. 1s. 6d.

8. *MAGNETISM;* being a concise Exposition of the General Principles of Magnetical Science. By Sir W. SNOW HARRIS. New Edition, revised by H. M. NOAD, Ph.D. With 165 Woodcuts. 3s. 6d.‡

☞ *The ‡ indicates that these vols. may be had strongly bound at 6d. extra.*

Physical Science, Natural Philosophy, etc., *continued.*

11. *THE ELECTRIC TELEGRAPH;* its History and Progress; with Descriptions of some of the Apparatus. By R. SABINE, C.E., F.S.A. 3s.

12. *PNEUMATICS,* for the Use of Beginners. By CHARLES TOMLINSON. Illustrated. 1s. 6d.

72. *MANUAL OF THE MOLLUSCA;* a Treatise on Recent and Fossil Shells. By Dr. S. P. WOODWARD, A.L.S. Fourth Edition. With Appendix by RALPH TATE, A.L.S., F.G.S. With numerous Plates and 300 Woodcuts. 6s. 6d. Cloth boards, 7s. 6d.

96. *ASTRONOMY.* By the late Rev. ROBERT MAIN, M.A. Third Edition, by WILLIAM THYNNE LYNN, B.A., F.R.A.S. 2s.

97. *STATICS AND DYNAMICS,* the Principles and Practice of; embracing also a clear development of Hydrostatics, Hydrodynamics, and Central Forces. By T. BAKER, C.E. 1s. 6d.

138. *TELEGRAPH,* Handbook of the; a Guide to Candidates for Employment in the Telegraph Service. By R. BOND. Fourth Edition. Including Questions on Magnetism, Electricity, and Practical Telegraphy, by W. McGREGOR. 3s.‡

173. *PHYSICAL GEOLOGY,* partly based on Major-General PORTLOCK'S "Rudiments of Geology." By RALPH TATE, A.L.S., &c. Woodcuts. 2s.

174. *HISTORICAL GEOLOGY,* partly based on Major-General PORTLOCK'S "Rudiments." By RALPH TATE, A.L.S., &c. Woodcuts. 2s. 6d.

173 & 174. *RUDIMENTARY TREATISE ON GEOLOGY,* Physical and Historical. Partly based on Major-General PORTLOCK'S "Rudiments of Geology." By RALPH TATE, A.L.S., F.G.S., &c. In One Volume. 4s. 6d.‡

183 & 184. *ANIMAL PHYSICS,* Handbook of. By Dr. LARDNER, D.C.L., formerly Professor of Natural Philosophy and Astronomy in University College, Lond. With 520 Illustrations. In One Vol. 7s. 6d., cloth boards.

*** *Sold also in Two Parts, as follows:—*

183. ANIMAL PHYSICS. By Dr. LARDNER. Part I., Chapters I.—VII. 4s.

184. ANIMAL PHYSICS. By Dr. LARDNER. Part II., Chapters VIII.—XVIII. 3s.

FINE ARTS.

20. *PERSPECTIVE FOR BEGINNERS.* Adapted to Young Students and Amateurs in Architecture, Painting, &c. By GEORGE PYNE. 2s.

40 *GLASS STAINING, AND THE ART OF PAINTING ON GLASS.* From the German of Dr. GESSERT and EMANUEL OTTO FROMBERG. With an Appendix on THE ART OF ENAMELLING. 2s. 6d.

69. *MUSIC,* A Rudimentary and Practical Treatise on. With numerous Examples. By CHARLES CHILD SPENCER. 2s. 6d.

71. *PIANOFORTE,* The Art of Playing the. With numerous Exercises & Lessons from the Best Masters. By CHARLES CHILD SPENCER. 1s. 6d.

69-71. *MUSIC & THE PIANOFORTE.* In one vol. Half bound, 5s.

181. *PAINTING POPULARLY EXPLAINED,* including Fresco, Oil, Mosaic, Water Colour, Water-Glass, Tempera, Encaustic, Miniature, Painting on Ivory, Vellum, Pottery, Enamel, Glass, &c. With Historical Sketches of the Progress of the Art by THOMAS JOHN GULLICK, assisted by JOHN TIMBS, F.S.A. Fifth Edition, revised and enlarged. 5s.‡

186. *A GRAMMAR OF COLOURING,* applied to Decorative Painting and the Arts. By GEORGE FIELD. New Edition, enlarged and adapted to the Use of the Ornamental Painter and Designer. By ELLIS A. DAVIDSON. With two new Coloured Diagrams, &c. 3s.‡

246. *A DICTIONARY OF PAINTERS, AND HANDBOOK FOR PICTURE AMATEURS;* including Methods of Painting, Cleaning, Relining and Restoring, Schools of Painting, &c. With Notes on the Copyists and Imitators of each Master. By PHILIPPE DARYL. 2s. 6d.‡

☞ *The ‡ indicates that these vols. may be had strongly bound at 6d. extra.*

INDUSTRIAL AND USEFUL ARTS.

23. *BRICKS AND TILES*, Rudimentary Treatise on the Manufacture of. By E. DOBSON, M.R.I.B.A. Illustrated, 3s.‡

67. *CLOCKS, WATCHES, AND BELLS*, a Rudimentary Treatise on. By SIR EDMUND BECKETT, LL.D., Q.C. Seventh Edition, revised and enlarged. 4s. 6d. limp; 5s. 6d. cloth boards.

83**. *CONSTRUCTION OF DOOR LOCKS.* Compiled from the Papers of A. C. HOBBS, and Edited by CHARLES TOMLINSON, F.R.S. With Additions by ROBERT MALLET, M.I.C.E. Illus. 2s. 6d.

162. *THE BRASS FOUNDER'S MANUAL;* Instructions for Modelling, Pattern-Making, Moulding, Turning, Filing, Burnishing, Bronzing, &c. With copious Receipts, &c. By WALTER GRAHAM. 2s.‡

205. *THE ART OF LETTER PAINTING MADE EASY.* By J. G. BADENOCH. Illustrated with 12 full-page Engravings of Examples. 1s.

215. *THE GOLDSMITH'S HANDBOOK*, containing full Instructions for the Alloying and Working of Gold. By GEORGE E. GEE, 3s.‡

225. *THE SILVERSMITH'S HANDBOOK*, containing full Instructions for the Alloying and Working of Silver. By GEORGE E. GEE. 3s.‡

*** *The two preceding Works, in One handsome Vol., half-bound, entitled* "THE GOLDSMITH'S & SILVERSMITH'S COMPLETE HANDBOOK," 7s. [*Just published.*

224. *COACH BUILDING*, A Practical Treatise, Historical and Descriptive. By J. W. BURGESS. 2s. 6d.‡

235. *PRACTICAL ORGAN BUILDING.* By W. E. DICKSON, M.A., Precentor of Ely Cathedral. Illustrated. 2s. 6d.‡

249. *THE HALL-MARKING OF JEWELLERY PRACTICALLY CONSIDERED.* By GEORGE E. GEE. 3s.‡

MISCELLANEOUS VOLUMES.

36. *A DICTIONARY OF TERMS used in ARCHITECTURE, BUILDING, ENGINEERING, MINING, METALLURGY, ARCHÆOLOGY, the FINE ARTS, &c.* By JOHN WEALE. Fifth Edition. Revised by ROBERT HUNT, F.R.S. Illustrated. 5s. limp; 6s. cloth boards.

50. *THE LAW OF CONTRACTS FOR WORKS AND SERVICES.* By DAVID GIBBONS. Third Edition, enlarged. 3s.‡

112. *MANUAL OF DOMESTIC MEDICINE.* By R. GOODING, B.A., M.D. A Family Guide in all Cases of Accident and Emergency. 2s.‡

112*. *MANAGEMENT OF HEALTH.* A Manual of Home and Personal Hygiene. By the Rev. JAMES BAIRD, B.A. 1s.

150. *LOGIC*, Pure and Applied. By S. H. EMMENS. 1s. 6d.

153. *SELECTIONS FROM LOCKE'S ESSAYS ON THE HUMAN UNDERSTANDING.* With Notes by S. H. EMMENS. 2s.

154. *GENERAL HINTS TO EMIGRANTS.* Notices of the various Fields for Emigration, Hints on Outfits, Useful Recipes, &c. 2s.

157. *THE EMIGRANT'S GUIDE TO NATAL.* By ROBERT JAMES MANN, F.R.A.S., F.M.S. Second Edition. Map. 2s.

193. *HANDBOOK OF FIELD FORTIFICATION*, intended for the Guidance of Officers Preparing for Promotion. By Major W. W. KNOLLYS, F.R.G.S. With 163 Woodcuts. 3s.‡

194. *THE HOUSE MANAGER:* Being a Guide to Housekeeping. Practical Cookery, Pickling and Preserving, Household Work, Dairy Management, the Table and Dessert, Cellarage of Wines, Home-brewing and Wine-making, the Boudoir and Dressing-room, Travelling, Stable Economy, Gardening Operations, &c. By AN OLD HOUSEKEEPER. 3s. 6d.‡

194, *HOUSE BOOK (The).* Comprising:—I. THE HOUSE MANAGER.
112 & By an OLD HOUSEKEEPER. II. DOMESTIC MEDICINE. By R. GOODING, M.D.
112*. III. MANAGEMENT OF HEALTH. By J. BAIRD. In One Vol., half-bound, 6s.

☞ *The ‡ indicates that these vols. may be had strongly bound at 6d. extra.*

EDUCATIONAL AND CLASSICAL SERIES.

HISTORY.

1. **England, Outlines of the History of;** more especially with reference to the Origin and Progress of the English Constitution. By WILLIAM DOUGLAS HAMILTON, F.S.A., of Her Majesty's Public Record Office. 4th Edition, revised. 5s.; cloth boards, 6s.

5. **Greece, Outlines of the History of;** in connection with the Rise of the Arts and Civilization in Europe. By W. DOUGLAS HAMILTON, of University College, London, and EDWARD LEVIEN, M.A., of Balliol College, Oxford. 2s. 6d.; cloth boards, 3s. 6d.

7. **Rome, Outlines of the History of:** from the Earliest Period to the Christian Era and the Commencement of the Decline of the Empire. By EDWARD LEVIEN, of Balliol College, Oxford. Map, 2s. 6d.; cl. bds. 3s. 6d.

9. **Chronology of History, Art, Literature, and Progress,** from the Creation of the World to the Present Time. The Continuation by W. D. HAMILTON, F.S.A. 3s.; cloth boards, 3s. 6d.

50. **Dates and Events in English History,** for the use of Candidates in Public and Private Examinations. By the Rev. E. RAND. 1s.

ENGLISH LANGUAGE AND MISCELLANEOUS.

11. **Grammar of the English Tongue,** Spoken and Written. With an Introduction to the Study of Comparative Philology. By HYDE CLARKE, D.C.L. Fourth Edition. 1s. 6d.

11*. **Philology:** Handbook of the Comparative Philology of English, Anglo-Saxon, Frisian, Flemish or Dutch, Low or Platt Dutch, High Dutch or German, Danish, Swedish, Icelandic, Latin, Italian, French, Spanish, and Portuguese Tongues. By HYDE CLARKE, D.C.L. 1s.

12. **Dictionary of the English Language,** as Spoken and Written. Containing above 100,000 Words. By HYDE CLARKE, D.C.L. 3s. 6d.; cloth boards, 4s. 6d.; complete with the GRAMMAR, cloth bds., 5s. 6d.

48. **Composition and Punctuation,** familiarly Explained for those who have neglected the Study of Grammar. By JUSTIN BRENAN. 17th Edition. 1s. 6d.

49. **Derivative Spelling-Book:** Giving the Origin of Every Word from the Greek, Latin, Saxon, German, Teutonic, Dutch, French, Spanish, and other Languages; with their present Acceptation and Pronunciation. By J. ROWBOTHAM, F.R.A.S. Improved Edition. 1s. 6d.

51. **The Art of Extempore Speaking:** Hints for the Pulpit, the Senate, and the Bar. By M. BAUTAIN, Vicar-General and Professor at the Sorbonne. Translated from the French. 8th Edition, carefully corrected. 2s. 6d.

52. **Mining and Quarrying,** with the Sciences connected therewith. First Book of, for Schools. By J. H. COLLINS, F.G.S., Lecturer to the Miners' Association of Cornwall and Devon. 1s.

53. **Places and Facts in Political and Physical Geography,** for Candidates in Examinations. By the Rev. EDGAR RAND, B.A. 1s.

54. **Analytical Chemistry,** Qualitative and Quantitative, a Course of. To which is prefixed, a Brief Treatise upon Modern Chemical Nomenclature and Notation. By WM. W. PINK and GEORGE E. WEBSTER. 2s.

THE SCHOOL MANAGERS' SERIES OF READING BOOKS,

Edited by the Rev. A. R. GRANT, Rector of Hitcham, and Honorary Canon of Ely; formerly H.M. Inspector of Schools.

INTRODUCTORY PRIMER, 3d.

	s.	d.		s.	d.
FIRST STANDARD . .	0	6	FOURTH STANDARD	1	2
SECOND " . .	0	10	FIFTH "	1	6
THIRD " . .	1	0	SIXTH "	1	6

LESSONS FROM THE BIBLE. Part I. Old Testament. 1s.

LESSONS FROM THE BIBLE. Part II. New Testament, to which is added THE GEOGRAPHY OF THE BIBLE, for very young Children. By Rev. C. THORNTON FORSTER. 1s. 2d. *** Or the Two Parts in One Volume. 2s.

FRENCH.

24. **French Grammar.** With Complete and Concise Rules on the Genders of French Nouns. By G. L. STRAUSS, Ph.D. 1s. 6d.

25. **French-English Dictionary.** Comprising a large number of New Terms used in Engineering, Mining, &c. By ALFRED ELWES. 1s. 6d.

26. **English-French Dictionary.** By ALFRED ELWES. 2s.

25,26. **French Dictionary** (as above). Complete, in One Vol., 3s.; cloth boards, 3s. 6d. *** Or with the GRAMMAR, cloth boards, 4s. 6d.

47. **French and English Phrase Book:** containing Introductory Lessons, with Translations, several Vocabularies of Words, a Collection of suitable Phrases, and Easy Familiar Dialogues. 1s. 6d.

GERMAN.

39. **German Grammar.** Adapted for English Students, from Heyse's Theoretical and Practical Grammar, by Dr. G. L. STRAUSS. 1s. 6d.

40. **German Reader:** A Series of Extracts, carefully culled from the most approved Authors of Germany; with Notes, Philological and Explanatory. By G. L. STRAUSS, Ph.D. 1s.

41-43. **German Triglot Dictionary.** By N. E. S. A. HAMILTON. In Three Parts. Part I. German-French-English. Part II. English-German-French. Part III. French-German-English. 3s., or cloth boards, 4s.

41-43 & 39. **German Triglot Dictionary** (as above), together with German Grammar (No. 39), in One Volume, cloth boards, 5s.

ITALIAN.

27. **Italian Grammar,** arranged in Twenty Lessons, with a Course of Exercises. By ALFRED ELWES. 1s. 6d.

28. **Italian Triglot Dictionary,** wherein the Genders of all the Italian and French Nouns are carefully noted down. By ALFRED ELWES. Vol. 1. Italian-English-French. 2s. 6d.

30. **Italian Triglot Dictionary.** By A. ELWES. Vol. 2. English-French-Italian. 2s. 6d.

32. **Italian Triglot Dictionary.** By ALFRED ELWES. Vol. 3. French-Italian-English. 2s. 6d.

28,30, 32. **Italian Triglot Dictionary** (as above). In One Vol., 7s. 6d. Cloth boards.

SPANISH AND PORTUGUESE.

34. **Spanish Grammar,** in a Simple and Practical Form. With a Course of Exercises. By ALFRED ELWES. 1s. 6d.

35. **Spanish-English and English-Spanish Dictionary.** Including a large number of Technical Terms used in Mining, Engineering, &c., with the proper Accents and the Gender of every Noun. By ALFRED ELWES. 4s.; cloth boards, 5s. *** Or with the GRAMMAR, cloth boards, 6s.

55. **Portuguese Grammar,** in a Simple and Practical Form. With a Course of Exercises. By ALFRED ELWES. 1s. 6d.

56. **Portuguese-English and English-Portuguese Dictionary.** Including a large number of Technical Terms used in Mining, Engineering, &c., with the proper Accents and the Gender of every Noun. By ALFRED ELWES. 5s.; cloth boards, 6s. *** Or with the GRAMMAR, cloth boards, 7s.

HEBREW.

46*. **Hebrew Grammar.** By Dr. BRESSLAU. 1s. 6d.

44. **Hebrew and English Dictionary,** Biblical and Rabbinical; containing the Hebrew and Chaldee Roots of the Old Testament Post-Rabbinical Writings. By Dr. BRESSLAU. 6s.

46. **English and Hebrew Dictionary.** By Dr. BRESSLAU. 3s.

44,46. 46*. **Hebrew Dictionary** (as above), in Two Vols., complete, with the GRAMMAR, cloth boards, 12s.

LATIN.

19. **Latin Grammar.** Containing the Inflections and Elementary Principles of Translation and Construction. By the Rev. THOMAS GOODWIN, M.A., Head Master of the Greenwich Proprietary School. 1s. 6d.

20. **Latin-English Dictionary.** By the Rev. THOMAS GOODWIN, M.A. 2s.

22. **English-Latin Dictionary;** together with an Appendix of French and Italian Words which have their origin from the Latin. By the Rev. THOMAS GOODWIN, M.A. 1s. 6d.

20,22. **Latin Dictionary** (as above). Complete in One Vol., 3s. 6d. cloth boards, 4s. 6d. *** Or with the GRAMMAR, cloth boards, 5s. 6d.

LATIN CLASSICS. With Explanatory Notes in English.

1. **Latin Delectus.** Containing Extracts from Classical Authors, with Genealogical Vocabularies and Explanatory Notes, by H. YOUNG. 1s. 6d.

2. **Cæsaris** Commentarii de Bello Gallico. Notes, and a Geographical Register for the Use of Schools, by H. YOUNG. 2s.

3. **Cornelius Nepos.** With Notes. By H. YOUNG. 1s.

4. **Virgilii** Maronis Bucolica et Georgica. With Notes on the Bucolics by W. RUSHTON, M.A., and on the Georgics by H. YOUNG. 1s. 6d.

5. **Virgilii** Maronis Æneis. With Notes, Critical and Explanatory, by H. YOUNG. New Edition, revised and improved With copious Additional Notes by Rev. T. H. L. LEARY, D.C.L., formerly Scholar of Brasenose College, Oxford. 3s.

5* ——— Part 1. Books i.—vi., 1s. 6d.

5** ——— Part 2. Books vii.—xii., 2s.

6. **Horace;** Odes, Epode, and Carmen Sæculare. Notes by H. YOUNG. 1s. 6d.

7. **Horace;** Satires, Epistles, and Ars Poetica. Notes by W. BROWNRIGG SMITH, M.A., F.R.G.S. 1s. 6d.

8. **Sallustii** Crispi Catalina et Bellum Jugurthinum. Notes, Critical and Explanatory, by W. M. DONNE, B.A., Trin. Coll., Cam. 1s. 6d.

9. **Terentii** Andria et Heautontimorumenos. With Notes, Critical and Explanatory, by the Rev. JAMES DAVIES, M.A. 1s. 6d.

10. **Terentii** Adelphi, Hecyra, Phormio. Edited, with Notes, Critical and Explanatory, by the Rev. JAMES DAVIES, M.A. 2s.

11. **Terentii** Eunuchus, Comœdia. Notes, by Rev. J. DAVIES, M.A. 1s. 6d.

12. **Ciceronis** Oratio pro Sexto Roscio Amerino. Edited, with an Introduction, Analysis, and Notes, Explanatory and Critical, by the Rev JAMES DAVIES, M.A. 1s. 6d.

13. **Ciceronis** Orationes in Catilinam, Verrem, et pro Archia. With Introduction, Analysis, and Notes, Explanatory and Critical, by Rev. T. H. L. LEARY, D.C.L. formerly Scholar of Brasenose College, Oxford. 1s. 6d.

14. **Ciceronis** Cato Major, Lælius, Brutus, sive de Senectute, de Amicitia, de Claris Oratoribus Dialogi. With Notes by W. BROWNRIGG SMITH, M.A., F.R.G.S. 2s.

16. **Livy:** History of Rome. Notes by H. YOUNG and W. B. SMITH, M.A. Part 1. Books i., ii., 1s. 6d.

16*. ——— Part 2. Books iii., iv., v., 1s. 6d.

17. ——— Part 3. Books xxi., xxii., 1s. 6d.

19. **Latin Verse Selections,** from Catullus, Tibullus, Propertius, and Ovid. Notes by W. B. DONNE, M.A., Trinity College, Cambridge. 2s.

20. **Latin Prose Selections,** from Varro, Columella, Vitruvius, Seneca, Quintilian, Florus, Velleius Paterculus, Valerius Maximus Suetonius, Apuleius, &c. Notes by W. B. DONNE, M.A. 2s.

21. **Juvenalis** Satiræ. With Prolegomena and Notes by T. H. S. ESCOTT, B.A., Lecturer on Logic at King's College, London. 2s.

GREEK.

14. **Greek Grammar,** in accordance with the Principles and Philological Researches of the most eminent Scholars of or own day. By HANS CLAUDE HAMILTON. 1s. 6d.

15,17. **Greek Lexicon.** Containing all the Words in General Use, with their Significations, Inflections, and Doubtful Quantities. By HENRY R. HAMILTON. Vol. 1. Greek-English, 2s. 6d.; Vol. 2. English-Greek, 2s. Or the Two Vols. in One, 4s. 6d.: cloth boards, 5s.

14,15, 17. **Greek Lexicon** (as above). Complete, with the GRAMMAR, in One Vol., cloth boards, 6s.

GREEK CLASSICS. With Explanatory Notes in English.

1. **Greek Delectus.** Containing Extracts from Classical Authors, with Genealogical Vocabularies and Explanatory Notes, by H. YOUNG. New Edition, with an improved and enlarged Supplementary Vocabulary, by JOHN HUTCHISON, M.A., of the High School, Glasgow. 1s. 6d.

2, 3. **Xenophon's Anabasis;** or, The Retreat of the Ten Thousand. Notes and a Geographical Register, by H. YOUNG. Part 1. Books i. to iii., 1s. Part 2. Books iv. to vii., 1s.

4. **Lucian's Select Dialogues.** The Text carefully revised, with Grammatical and Explanatory Notes, by H. YOUNG. 1s. 6d.

5-12. **Homer, The Works of.** According to the Text of BAEUMLEIN. With Notes, Critical and Explanatory, drawn from the best and latest Authorities, with Preliminary Observations and Appendices, by T. H. L. LEARY, M.A., D.C.L.

THE ILIAD:	Part 1. Books i. to vi., 1s. 6d.	Part 3. Books xiii. to xviii., 1s. 6d.
	Part 2. Books vii. to xii., 1s. 6d.	Part 4. Books xix. to xxiv., 1s. 6d.
THE ODYSSEY:	Part 1. Books i. to vi., 1s. 6d	Part 3. Books xiii. to xviii., 1s. 6d.
	Part 2. Books vii. to xii., 1s. 6d.	Part 4. Books xix. to xxiv., and Hymns, 2s.

13. **Plato's Dialogues:** The Apology of Socrates, the Crito, and the Phædo. From the Text of C. F. HERMANN. Edited with Notes, Critical and Explanatory, by the Rev. JAMES DAVIES, M.A. 2s.

14-17. **Herodotus,** The History of, chiefly after the Text of GAISFORD. With Preliminary Observations and Appendices, and Notes, Critical and Explanatory, by T. H. L. LEARY, M.A., D.C.L.
Part 1. Books i., ii. (The Clio and Euterpe), 2s.
Part 2. Books iii., iv. (The Thalia and Melpomene), 2s.
Part 3. Books v.-vii. (The Terpsichore, Erato, and Polymnia), 2s.
Part 4. Books viii., ix. (The Urania and Calliope) and Index, 1s. 6d.

18. **Sophocles:** Œdipus Tyrannus. Notes by H. YOUNG. 1s.

20. **Sophocles:** Antigone. From the Text of DINDORF. Notes, Critical and Explanatory, by the Rev. JOHN MILNER, B.A. 2s.

23. **Euripides:** Hecuba and Medea. Chiefly from the Text of DINDORF. With Notes, Critical and Explanatory, by W. BROWNRIGG SMITH, M.A., F.R.G.S. 1s. 6d.

26. **Euripides:** Alcestis. Chiefly from the Text of DINDORF. With Notes, Critical and Explanatory, by JOHN MILNER, B.A. 1s. 6d.

30. **Æschylus:** *Prometheus Vinctus*: The *Prometheus* Bound. From the Text of DINDORF. Edited, with English Notes, Critical and Explanatory, by the Rev. JAMES DAVIES, M.A. 1s.

32. **Æschylus:** Septem Contra Thebes: The Seven against Thebes. From the Text of DINDORF. Edited, with English Notes, Critical and Explanatory, by the Rev. JAMES DAVIES, M.A. 1s.

40. **Aristophanes:** Acharnians. Chiefly from the Text of C. H. WEISE. With Notes, by C. S. T. TOWNSHEND, M.A. 1s. 6d.

41. **Thucydides:** History of the Peloponnesian War. Notes by H. YOUNG. Book 1. 1s. 6d.

42. **Xenophon's Panegyric on Agesilaus.** Notes and Introduction by LL. F. W. JEWITT. 1s. 6d.

43. **Demosthenes.** The Oration on the Crown and the Philippics. With English Notes. By Rev. T. H. L. LEARY, D.C.L., formerly Scholar of Brasenose College, Oxford. 1s. 6d.

7, STATIONERS' HALL COURT, LONDON, E.C.
October, 1886.

A CATALOGUE OF BOOKS

INCLUDING MANY NEW AND STANDARD WORKS IN

ENGINEERING, ARCHITECTURE, MECHANICS, MATHEMATICS, SCIENCE, THE INDUSTRIAL ARTS, AGRICULTURE, LAND MANAGEMENT, GARDENING, &c.

PUBLISHED BY

CROSBY LOCKWOOD & CO.

CIVIL ENGINEERING, SURVEYING, etc.

The Water Supply of Cities and Towns.

A COMPREHENSIVE TREATISE on the WATER-SUPPLY OF CITIES AND TOWNS. By WILLIAM HUMBER, A-M. Inst. C.E., and M. Inst. M.E., Author of "Cast and Wrought Iron Bridge Construction," &c., &c. Illustrated with 50 Double Plates, 1 Single Plate, Coloured Frontispiece, and upwards of 250 Woodcuts, and containing 400 pages of Text. Imp. 4to, £6 6s. elegantly and substantially half-bound in morocco.

List of Contents.

I. Historical Sketch of some of the means that have been adopted for the Supply of Water to Cities and Towns.—II. Water and the Foreign Matter usually associated with it.—III. Rainfall and Evaporation.—IV. Springs and the water-bearing formations of various districts.—V. Measurement and Estimation of the flow of Water.—VI. On the Selection of the Source of Supply.—VII. Wells.—VIII. Reservoirs.—IX. The Purification of Water.—X. Pumps.—XI. Pumping Machinery.—XII. Conduits.—XIII. Distribution of Water.—XIV Meters, Service Pipes, and House Fittings.—XV. The Law and Economy of Water Works. XVI. Constant and Intermittent Supply.—XVII. Description of Plates.—Appendices, giving Tables of Rates of Supply, Velocities, &c. &c., together with Specifications of several Works illustrated, among which will be found: Aberdeen, Bideford, Canterbury, Dundee, Halifax, Lambeth, Rotherham, Dublin, and others.

"The most systematic and valuable work upon water supply hitherto produced in English, or in any other language. . . . Mr. Humber's work is characterised almost throughout by an exhaustiveness much more distinctive of French and German than of English technical treatises."—*Engineer.*

"We can congratulate Mr. Humber on having been able to give so large an amount of information on a subject so important as the water supply of cities and towns. The plates, fifty in number, are mostly drawings of executed works, and alone would have commanded the attention of every engineer whose practice may lie in this branch of the profession."—*Builder.*

Cast and Wrought Iron Bridge Construction.

A COMPLETE AND PRACTICAL TREATISE ON CAST AND WROUGHT IRON BRIDGE CONSTRUCTION, including Iron Foundations. In Three Parts—Theoretical, Practical, and Descriptive. By WILLIAM HUMBER, A-M. Inst. C.E., and M. Inst. M.E. Third Edition, Revised and much improved, with 115 Double Plates (20 of which now first appear in this edition), and numerous Additions to the Text. In Two Vols., imp. 4to, £6 16s. 6d. half-bound in morocco.

"A very valuable contribution to the standard literature of civil engineering. In addition to elevations, plans and sections, large scale details are given which very much enhance the instructive worth of these illustrations. No engineer would willingly be without so valuable a fund of information."—*Civil Engineer and Architect's Journal.*

"Mr. Humber's stately volumes, lately issued—in which the most important bridges erected during the last five years, under the direction of the late Mr. Brunel, Sir W. Cubitt, Mr. Hawkshaw, Mr. Page, Mr. Fowler, Mr. Hemans, and others among our most eminent engineers, are drawn and specified in great detail."—*Engineer.*

HUMBER'S GREAT WORK ON MODERN ENGINEERING.

Complete in Four Volumes, imperial 4to, price £12 12s., half-morocco, each Volume sold separately as follows:—

A RECORD OF THE PROGRESS OF MODERN ENGINEERING. FIRST SERIES. Comprising Civil, Mechanical, Marine, Hydraulic, Railway, Bridge, and other Engineering Works, &c. By WILLIAM HUMBER, A-M. Inst. C.E., &c. Imp. 4to, with 36 Double Plates, drawn to a large scale, Photographic Portrait of John Hawkshaw, C.E., F.R.S., &c., and copious descriptive Letterpress, Specifications, &c., £3 3s. half-morocco.

List of the Plates and Diagrams.

Victoria Station and Roof, L. B. & S. C. R. (8 plates); Southport Pier (2 plates); Victoria Station and Roof, L. C. & D. and G. W. R. (6 plates); Roof of Cremorne Music Hall; Bridge over G. N. Railway; Roof of Station, Dutch Rhenish Rail (2 plates); Bridge over the Thames, West London Extension Railway (5 plates); Armour Plates: Suspension Bridge, Thames (4 plates); The Allen Engine; Suspension Bridge, Avon (3 plates); Underground Railway (3 plates).

"Handsomely lithographed and printed. It will find favour with many who desire to preserve in a permanent form copies of the plans and specifications prepared for the guidance of the contractors for many important engineering works."—*Engineer.*

HUMBER'S RECORD OF MODERN ENGINEERING. SECOND SERIES. Imp. 4to, with 36 Double Plates, Photographic Portrait of Robert Stephenson, C.E., M.P., F.R.S., &c., and copious descriptive Letterpress, Specifications, &c., £3 3s. half-morocco.

List of the Plates and Diagrams.

Birkenhead Docks, Low Water Basin (15 plates); Charing Cross Station Roof, C. C. Railway (3 plates); Digswell Viaduct, Great Northern Railway; Robbery Wood Viaduct, Great Northern Railway; Iron Permanent Way; Clydach Viaduct, Merthyr, Tredegar, and Abergavenny Railway; Ebbw Viaduct, Merthyr, Tredegar, and Abergavenny Railway; College Wood Viaduct, Cornwall Railway; Dublin Winter Palace Roof (3 plates); Bridge over the Thames, L. C. & D. Railway (6 plates); Albert Harbour, Greenock (4 plates).

"Mr. Humber has done the profession good and true service, by the fine collection of examples he has here brought before the profession and the public."—*Practical Mechanic's Journal.*

HUMBER'S RECORD OF MODERN ENGINEERING. THIRD SERIES. Imp. 4to, with 40 Double Plates, Photographic Portrait of J. R. M'Clean, Esq., late Pres. Inst. C.E., and copious descriptive Letterpress, Specifications, &c., £3 3s. half-morocco.

List of the Plates and Diagrams.

MAIN DRAINAGE, METROPOLIS.—*North Side.*—Map showing Interception of Sewers; Middle Level Sewer (2 plates); Outfall Sewer, Bridge over River Lea (3 plates); Outfall Sewer, Bridge over Marsh Lane, North Woolwich Railway, and Bow and Barking Railway Junction; Outfall Sewer, Bridge over Bow and Barking Railway (3 plates); Outfall Sewer, Bridge over East London Waterworks' Feeder (2 plates); Outfall Sewer, Reservoir (2 plates); Outfall Sewer, Tumbling Bay and Outlet; Outfall Sewer, Penstocks. *South Side.*—Outfall Sewer, Bermondsey Branch (2 plates); Outfall Sewer, Reservoir and Outlet (4 plates); Outfall Sewer, Filth Hoist; Sections of Sewers (North and South Sides).

THAMES EMBANKMENT.—Section of River Wall; Steamboat Pier, Westminster (2 plates); Landing Stairs between Charing Cross and Waterloo Bridges; York Gate (2 plates); Overflow and Outlet at Savoy Street Sewer (3 plates); Steamboat Pier, Waterloo Bridge (3 plates); Junction of Sewers, Plans and Sections; Gullies, Plans and Sections; Rolling Stock; Granite and Iron Forts.

"The drawings have a constantly increasing value, and whoever desires to possess clear representations of the two great works carried out by our Metropolitan Board will obtain Mr. Humber's volume."—*Engineer.*

HUMBER'S RECORD OF MODERN ENGINEERING. FOURTH SERIES. Imp. 4to, with 36 Double Plates, Photographic Portrait of John Fowler, Esq., late Pres. Inst. C.E., and copious descriptive Letterpress, Specifications, &c., £3 3s. half-morocco.

List of the Plates and Diagrams.

Abbey Mills Pumping Station, Main Drainage, Metropolis (4 plates); Barrow Docks (5 plates); Manquis Viaduct, Santiago and Valparaiso Railway (2 plates); Adam's Locomotive, St. Helen's Canal Railway (2 plates); Cannon Street Station Roof, Charing Cross Railway (3 plates); Road Bridge over the River Moka (2 plates); Telegraphic Apparatus for Mesopotamia; Viaduct over the River Wye, Midland Railway (3 plates); St. Germans Viaduct, Cornwall Railway (2 plates); Wrought-Iron Cylinder for Diving Bell; Millwall Docks (6 plates); Milroy's Patent Excavator; Metropolitan District Railway (6 plates); Harbours, Ports, and Breakwaters (3 plates).

"We gladly welcome another year's issue of this valuable publication from the able pen of Mr. Humber. The accuracy and general excellence of this work are well known, while its usefulness in giving the measurements and details of some of the latest examples of engineering, as carried out by the most eminent men in the profession, cannot be too highly prized."—*Artizan.*

Trigonometrical Surveying.

AN OUTLINE OF THE METHOD OF CONDUCTING A TRIGONOMETRICAL SURVEY, for the Formation of Geographical and Topographical Maps and Plans, Military Reconnaissance, Levelling, &c., with Useful Problems, Formulæ, and Tables. By Lieut.-General FROME, R.E. Fourth Edition, Revised and partly Re-written by Captain CHARLES WARREN, R.E. With 19 Plates and 115 Woodcuts, royal 8vo, 16s. cloth.

"The simple fact that a fourth edition has been called for is the best testimony to its merits. No words of praise from us can strengthen the position so well and so steadily maintained by this work. Captain Warren has revised the entire work, and made such additions as were necessary to bring every portion of the contents up to the present date."—*Broad Arrow.*

Oblique Bridges.

A PRACTICAL AND THEORETICAL ESSAY ON OBLIQUE BRIDGES. With 13 large Plates. By the late GEORGE WATSON BUCK, M.I.C.E. Third Edition, revised by his Son, J. H. WATSON BUCK, M.I.C.E. and with the addition of Description to Diagrams for Facilitating the Construction of Oblique Bridges by W. H. BARLOW, M.I.C.E. Royal 8vo, 12s. cloth.

"The standard text-book for all engineers regarding skew arches is Mr. Buck's treatise, and it would be impossible to consult a better."—*Engineer.*

"Mr. Buck's treatise is recognised as a standard text-book, and his treatment has divested the subject of many of the intricacies supposed to belong to it. As a guide to the engineer and architect, on a confessedly difficult subject, Mr. Buck's work is unsurpassed."—*Building News.*

Bridge Construction in Masonry, Timber and Iron.

EXAMPLES OF BRIDGE AND VIADUCT CONSTRUCTION OF MASONRY, TIMBER, AND IRON. Consisting of 46 Plates from the Contract Drawings or Admeasurement of select Works. By W. D. HASKOLL, C.E. Second Edition, with the addition of 554 Estimates, and the Practice of Setting out Works. Illustrated with 6 pages of Diagrams. Imp. 4to, £2 12s. 6*d.* half-morocco.

"A work of the present nature by a man of Mr. Haskoll's experience must prove invaluable. The tables of estimates will considerably enhance its value."—*Engineering.*

Earthwork.

EARTHWORK TABLES. Showing the Contents in Cubic Yards of Embankments, Cuttings, &c., of Heights or Depths up to an average of 80 feet. By JOSEPH BROADBENT, C.E., and FRANCIS CAMPIN, C.E. Crown 8vo, 5s. cloth.

"The way in which accuracy is attained, by a simple division of each cross section into three elements, two in which are constant and one variable, is ingenious."—*Athenæum.*

Barlow's Strength of Materials, enlarged.

A TREATISE ON THE STRENGTH OF MATERIALS; with Rules for Application in Architecture, the Construction of Suspension Bridges, Railways, &c. By PETER BARLOW, F.R.S. A New Edition, revised by his Sons, P. W. BARLOW, F.R.S., and W. H. BARLOW, F.R.S.; to which are added, Experiments by HODGKINSON, FAIRBAIRN, and KIRKALDY; and Formulæ for Calculating Girders, &c. Arranged and Edited by W. HUMBER, A-M. Inst. C.E. Demy 8vo, 400 pp., with 19 large Plates and numerous Woodcuts, 18s. cloth.

'Valuable alike to the student, tyro, and the experienced practitioner, it will always rank in future as it has hitherto done, as the standard treatise on that particular subject."—*Engineer.*

There is no greater authority than Barlow."—*Building News.*

"As a scientific work of the first class, it deserves a foremost place on the bookshelves of every civil engineer and practical mechanic."—*English Mechanic.*

Strains, Formulæ and Diagrams for Calculation of.

A HANDY BOOK FOR THE CALCULATION OF STRAINS IN GIRDERS AND SIMILAR STRUCTURES, AND THEIR STRENGTH. Consisting of Formulæ and Corresponding Diagrams, with numerous details for Practical Application, &c. By WILLIAM HUMBER, A-M. Inst. C.E., &c. Fourth Edition. Crown 8vo, nearly 100 Woodcuts and 3 Plates, 7s. 6*d.* cloth.

The formulæ are neatly expressed, and the diagrams good."—*Athenæum.*

"We heartily commend this really *handy* book to our engineer and architect readers."—*English Mechanic.*

Survey Practice.

AID TO SURVEY PRACTICE, for Reference in Surveying, Levelling, Setting-out and in Route Surveys of Travellers by Land and Sea. With Tables, Illustrations, and Records. By LOWIS D'A. JACKSON, A.M.I.C.E., Author of "Hydraulic Manual," "Modern Metrology," &c. Large crown 8vo, 12s. 6d. cloth.

"Mr. Jackson has produced a valuable *vade-mecum* for the surveyor. We can recommend this book as containing an admirable supplement to the teaching of the accomplished surveyor."—*Athenæum.*

"As a text-book we should advise all surveyors to place it in their libraries, and study well the matured instructions afforded in its pages."—*Colliery Guardian.*

"The author brings to his work a fortunate union of theory and practical experience which, aided by a clear and lucid style of writing, renders the book a very useful one."—*Builder.*

Surveying, Land and Marine.

LAND AND MARINE SURVEYING, in Reference to the Preparation of Plans for Roads and Railways; Canals, Rivers, Towns' Water Supplies: Docks and Harbours. With Description and Use of Surveying Instruments. By W. DAVIS HASKOLL, C.E., Author of "Bridge and Viaduct Construction," &c. Second Edition, Revised, with Additions. Large crown 8vo, 9s. cloth.

"This book must prove of great value to the student. We have no hesitation in recommending it, feeling assured that it will more than repay a careful study."—*Mechanical World.*

"A most useful and well arranged book for the aid of a student. We can strongly recommend it as a carefully written and valuable text-book. It enjoys a well-deserved repute among surveyors."—*Builder.*

Levelling.

A TREATISE ON THE PRINCIPLES AND PRACTICE OF LEVELLING. Showing its Application to purposes of Railway and Civil Engineering, in the Construction of Roads; with MR. TELFORD'S Rules for the same. By FREDERICK W. SIMMS, F.G.S., M. Inst. C.E. Seventh Edition, with the addition of LAW'S Practical Examples for Setting-out Railway Curves, and TRAUTWINE'S Field Practice of Laying-out Circular Curves. With 7 Plates and numerous Woodcuts, 8vo, 8s. 6d. cloth. *** TRAUTWINE on Curves, separate, 5s.

"The text-book on levelling in most of our engineering schools and colleges."—*Engineer.*

"The publishers have rendered a substantial service to the profession, especially to the younger members, by bringing out the present edition of Mr. Simms' useful work."—*Engineering.*

Tunnelling.

PRACTICAL TUNNELLING. Explaining in detail the Setting-out of the works, Shaft-sinking and Heading-driving, Ranging the Lines and Levelling underground, Sub-Excavating, Timbering, and the Construction of the Brickwork of Tunnels, with the amount of Labour required for, and the Cost of, the various portions of the work. By FREDERICK W. SIMMS, F.G.S., M. Inst. C.E. Third Edition, Revised and Extended by D. KINNEAR CLARK, M. Inst. C.E. Imp. 8vo, with 21 Folding Plates and numerous Wood Engravings, 30s. cloth.

"The estimation in which Mr. Simms' book on tunnelling has been held for over thirty years cannot be more truly expressed than in the words of the late Professor Rankine:—'The best source of information on the subject of tunnels is Mr. F. W. Simms' work on Practical Tunnelling.'"—*Architect.*

"It has been regarded from the first as a text-book of the subject. . . . . Mr. Clark has added immensely to the value of the book."—*Engineer.*

"The additional chapters by Mr. Clark, containing as they do numerous examples of modern practice, bring the book well up to date."—*Engineering.*

Statics, Graphic and Analytic.

GRAPHIC AND ANALYTIC STATICS, in Theory and Comparison: Their Practical Application to the Treatment of Stresses in Roofs, Solid Girders, Lattice, Bowstring and Suspension Bridges, Braced Iron Arches and Piers, and other Frameworks. To which is added a Chapter on Wind Pressures. By R. HUDSON GRAHAM, C.E. With numerous Examples, many taken from existing Structures. 8vo, 16s. cloth.

"Mr. Graham's book will find a place wherever graphic and analytic statics are used or studied."—*Engineer.*

"This exhaustive treatise is admirably adapted for the architect and engineer, and will tend to wean the profession from a tedious and laboured mode of calculation. To prove the accuracy of the graphical demonstrations, the author compares them with the analytic formulæ given by Rankine."—*Building News.*

"The work is excellent from a practical point of view, and has evidently been prepared with much care. It is an excellent text-book for the practical draughtsman."—*Athenæum.*

Hydraulic Tables.

HYDRAULIC TABLES, CO-EFFICIENTS, and FORMULÆ for finding the Discharge of Water from Orifices, Notches, Weirs, Pipes, and Rivers. With New Formulæ, Tables and General Information on Rainfall, Catchment-Basins, Drainage, Sewerage, Water Supply for Towns and Mill Power. By JOHN NEVILLE, Civil Engineer, M.R.I.A. Third Edition, carefully revised, with considerable Additions. Numerous Illustrations. Crown 8vo, 14s. cloth.

Alike valuable to students and engineers in practice; its study will prevent the annoyance of avoidable failures, and assist them to select the readiest means of successfully carrying out any given work connected with hydraulic engineering."—*Mining Journal.*

"It is, of all English books on the subject, the one nearest to completion. . . . From the good arrangement of the matter, the clear explanations, and abundance of formulæ, the carefully calculated tables, and, above all, the thorough acquaintance with both theory and construction which is displayed from first to last, the book will be found to be an acquisition."—*Architect.*

River Engineering.

RIVER BARS: The Causes of their Formation, and their Treatment by "Induced Tidal Scour." With a Description of the Successful Reduction by this Method of the Bar at Dublin. By I. J. MANN, Assist. Eng. to the Dublin Port and Docks Board. Royal 8vo, 7s. 6d. cloth.

"We recommend all interested in harbour works—and, indeed, those concerned in the improvements of rivers generally—to read Mr. Mann's interesting work on the treatment of river bars."—*Engineer.*

"The author's discussion on wave-action, currents, and scour is intelligent and interesting. . . a most valuable contribution to the history of this branch of engineering."—*Engineering and Mining Journal.*

Hydraulics.

HYDRAULIC MANUAL. Consisting of Working Tables and Explanatory Text. Intended as a Guide in Hydraulic Calculations and Field Operations. By LOWIS D'A. JACKSON. Fourth Edition. Rewritten and Enlarged. Large crown 8vo, 16s. cloth.

"From the great mass of material at his command the author has constructed a manual which may be accepted as a trustworthy guide to this branch of the engineer's profession. We can heartily recommend this volume to all who desire to be acquainted with the latest development of this important subject."—*Engineering.*

"The standard work in this department of mechanics. The present edition has been brought abreast of the most recent practice."—*Scotsman.*

"The most useful feature of this work is its freedom from what is superannuated and its thorough adoption of recent experiments; the text is, in fact, in great part a short account of the great modern experiments."—*Nature.*

Tramways and their Working.

TRAMWAYS: THEIR CONSTRUCTION AND WORKING. Embracing a Comprehensive History of the System; with an exhaustive Analysis of the various Modes of Traction, including Horse-Power, Steam, Heated Water, and Compressed Air; a Description of the Varieties of Rolling Stock; and ample Details of Cost and Working Expenses: the Progress recently made in Tramway Construction, &c. &c. By D. KINNEAR CLARK, M. Inst. C.E. With over 200 Wood Engravings, and 13 Folding Plates. Two Vols., large crown 8vo, 30s. cloth.

"All interested in tramways must refer to it, as all railway engineers have turned to the author's work 'Railway Machinery.'"—*Engineer.*

"An exhaustive and practical work on tramways, in which the history of this kind of locomotion, and a description and cost of the various modes of laying tramways, are to be found."—*Building News.*

"The best form of rails, the best mode of construction, and the best mechanical appliances are so fairly indicated in the work under review, that any engineer about to construct a tramway will be enabled at once to obtain the practical information which will be of most service to him."—*Athenæum.*

Oblique Arches.

A PRACTICAL TREATISE ON THE CONSTRUCTION OF OBLIQUE ARCHES. By JOHN HART. Third Edition, with Plates. Imperial 8vo, 8s. cloth.

Strength of Girders.

GRAPHIC TABLE FOR FACILITATING THE COMPUTATION OF THE WEIGHTS OF WROUGHT IRON AND STEEL GIRDERS, &c., for Parliamentary and other Estimates. By J. H. WATSON BUCK, M. Inst. C.E. On a Sheet, 2s. 6d.

Tables for Setting-out Curves.

TABLES OF TANGENTIAL ANGLES AND MULTIPLES for Setting-out Curves from 5 to 200 Radius. By ALEXANDER BEAZELEY, M. Inst. C.E. Third Edition. Printed on 48 Cards, and sold in a cloth box, waistcoat-pocket size, 3s. 6d.

"Each table is printed on a small card, which, being placed on the theodolite, leaves the hands free to manipulate the instrument—no small advantage as regards the rapidity of work."—*Engineer.*

"Very handy; a man may know that all his day's work must fall on two of these cards, which he puts into his own card-case, and leaves the rest behind."—*Athenæum.*

Engineering Fieldwork.

THE PRACTICE OF ENGINEERING FIELDWORK, applied to Land and Hydraulic, Hydrographic, and Submarine Surveying and Levelling. Second Edition, Revised, with considerable Additions, and a Supplement on Waterworks, Sewers, Sewage, and Irrigation. By W. DAVIS HASKOLL, C.E. Numerous Folding Plates. In One Volume, demy 8vo, £1 5s. cloth.

Large Tunnel Shafts.

THE CONSTRUCTION OF LARGE TUNNEL SHAFTS: A Practical and Theoretical Essay. By J. H. WATSON BUCK, M. Inst. C.E., Resident Engineer, London and North-Western Railway. Illustrated with Folding Plates, royal 8vo, 12s. cloth.

"Many of the methods given are of extreme practical value to the mason; and the observations on the form of arch, the rules for ordering the stone, and the construction of the templates will be found of considerable use. We commend the book to the engineering profession."—*Building News.*

"Will be regarded by civil engineers as of the utmost value, and calculated to save much time and obviate many mistakes."—*Colliery Guardian.*

Field-Book for Engineers.

THE ENGINEER'S, MINING SURVEYOR'S, AND CONTRACTOR'S FIELD-BOOK. Consisting of a Series of Tables, with Rules, Explanations of Systems, and use of Theodolite for Traverse Surveying and Plotting the Work with minute accuracy by means of Straight Edge and Set Square only; Levelling with the Theodolite, Casting-out and Reducing Levels to Datum, and Plotting Sections in the ordinary manner; setting-out Curves with the Theodolite by Tangential Angles and Multiples, with Right and Left-hand Readings of the Instrument: Setting-out Curves without Theodolite, on the System of Tangential Angles by sets of Tangents and Offsets: and Earthwork Tables to 80 feet deep, calculated for every 6 inches in depth. By W. DAVIS HASKOLL, C.E. With numerous Woodcuts. Fourth Edition, Enlarged. Crown 8vo, 12s. cloth.

"The book is very handy, and the author might have added that the separate tables of sines and tangents to every minute will make it useful for many other purposes, the genuine traverse tables existing all the same."—*Athenæum.*

"Every person engaged in engineering field operations will estimate the importance of such a work and the amount of valuable time which will be saved by reference to a set of reliable tables prepared with the accuracy and fulness of those given in this volume."—*Railway News.*

Earthwork, Measurement and Calculation of.

A MANUAL ON EARTHWORK. By ALEX. J. S. GRAHAM, C.E. With numerous Diagrams. 18mo, 2s. 6d. cloth.

"A great amount of practical information, very admirably arranged, and available for rough estimates, as well as for the more exact calculations required in the engineer's and contractor's offices."—*Artizan.*

Strains.

THE STRAINS ON STRUCTURES OF IRONWORK; with Practical Remarks on Iron Construction. By F. W. SHEILDS, M. Inst. C.E. Second Edition, with 5 Plates. Royal 8vo, 5s. cloth.

"The student cannot find a better little book on this subject."—*Engineer.*

Strength of Cast Iron, etc.

A PRACTICAL ESSAY ON THE STRENGTH OF CAST IRON AND OTHER METALS. By THOMAS TREDGOLD, C.E. Fifth Edition, including HODGKINSON'S Experimental Researches. 8vo, 12s. cloth.

MECHANICS & MECHANICAL ENGINEERING.

The Modernised "Templeton."

THE PRACTICAL MECHANIC'S WORKSHOP COMPANION. Comprising a great variety of the most useful Rules and Formulæ in Mechanical Science, with numerous Tables of Practical Data and Calculated Results for Facilitating Mechanical Operations. By WILLIAM TEMPLETON, Author of "The Engineer's Practical Assistant," &c. &c. An Entirely New Edition, Revised, Modernised, and considerably Enlarged by WALTER S. HUTTON, C.E., Author of "The Works' Manager's Handbook of Modern Rules, Tables, and Data for Engineers," &c. Fcap. 8vo, nearly 500 pp., with 8 Plates and upwards of 250 Illustrative Diagrams, 6s., strongly bound for workshop or pocket wear and tear. [*Just published.*

☞ TEMPLETON'S "MECHANIC'S WORKSHOP COMPANION" *has been for more than a quarter of a century deservedly popular, having run through numerous Editions; and, as a recognised Text-Book and well-worn and thumb-marked* vade mecum *of several generations of intelligent and aspiring workmen, it has had the reputation of having been the means of raising many of them in their position in life.*

In its present greatly Enlarged, *Improved and Modernised form, the Publishers are sure that it will commend itself to the English workmen of the present day all the world over, and become, like its predecessors, their indispensable friend and referee.*

A smaller type having been adopted, and the page increased in size, while the number of pages has advanced from about 330 *to nearly* 500, *the book practically contains double the amount of matter that was comprised in the original work.*

*** OPINIONS OF THE PRESS.

"In its modernised form Hutton's 'Templeton' should have a wide sale, fo it contains much valuable information which the mechanic will often find of use, and not a few tables and notes which he might look for in vain in other works. This modernised edition will be appreciated by all who have learned to value the original editions of 'Templeton.'"—*English Mechanic.*

"It has met with great success in the engineering workshop, as we can testify; and there are great many men who, in a great measure, owe their rise in life to this little book."—*Building ews.*

Engineer's and Machinist's Assistant.

THE ENGINEER'S, MILLWRIGHT'S, and MACHINIST'S PRACTICAL ASSISTANT. A collection of Useful Tables, Rules and Data. By WILLIAM TEMPLETON. Seventh Edition, with Additions. 18mo, 2s. 6d. cloth.

"Templeton's handbook occupies a foremost place among books of this kind. A more suitable esent to an apprentice to any of the mechanical trades could not possibly be made."—*Building ws.*

Turning.

LATHE-WORK: A Practical Treatise on the Tools, Appliances, and Processes employed in the Art of Turning. By PAUL N. HASLUCK. Third Edition, Revised and Enlarged. Crown 8vo, 5s. cloth.

"Written by a man who knows, not only how work ought to be done, but who also knows how do it, and how to convey his knowledge to others."—*Engineering.*

We can safely recommend the work to young engineers. To the amateur it will simply be valuable. To the student it will convey a great deal of useful information."—*Engineer.*

"A compact, succinct, and handy guide to lathe-work did not exist in our language until Mr. Hasluck, by the publication of this treatise, gave the turner a true *vade-mecum.*"—*House Decorator.*

Iron and Steel.

"IRON AND STEEL": A Work for the Forge, Foundry, Factory, and Office. Containing ready, useful, and trustworthy Information for Ironmasters and their Stock-takers; Managers of Bar, Rail, Plate, and Sheet Rolling Mills; Iron and Metal Founders; Iron Ship and Bridge Builders; Mechanical, Mining, and Consulting Engineers; Architects, Contractors, Builders, and Professional Draughtsmen. By CHARLES HOARE, Author of "The Slide Rule," &c. Eighth Edition, Revised throughout and considerably Enlarged. With folding Scales of "Foreign Measures compared with the English Foot," and "Fixed Scales of Squares, Cubes, and Roots, Areas, Decimal Equivalents, &c." Oblong 32mo, leather, elastic band, 6s.

"For comprehensiveness the book has not its equal."—*Iron.*

"One of the best of the pocket books, and a useful companion in other branches of work than iron and steel."—*English Mechanic.*

"We cordially recommend this book to those engaged in considering the details of all kinds of iron and steel works."—*Naval Science.*

Stone-working Machinery.

STONE-WORKING MACHINERY, and the Rapid and Economical Conversion of Stone. With Hints on the Arrangement and Management of Stone Works. By M. POWIS BALE, M.I.M.E., A.M.I.C.E. With numerous Illustrations. Large crown 8vo, 9*s.* cloth.

"The book should be in the hands of every mason or student of stone-work."—*Colliery Guardian.*

Engineer's Reference Book.

THE WORKS' MANAGER'S HANDBOOK OF MODERN RULES, TABLES, AND DATA. For Engineers, Millwrights, and Boiler Makers; Tool Makers, Machinists, and Metal Workers; Iron and Brass Founders, &c. By W. S. HUTTON, Civil and Mechanical Engineer. Third Edition, carefully revised, with Additions. In One handsome Volume, medium 8vo, price 15*s.* strongly bound.

"The author treats every subject from the point of view of one who has collected workshop notes for application in workshop practice, rather than from the theoretical or literary aspect. The volume contains a great deal of that kind of information which is gained only by practical experience, and is seldom written in books."—*Engineer.*

"The volume is an exceedingly useful one, brimful with engineers notes, memoranda, and rules, and well worthy of being on every mechanical engineer's bookshelf. . . . There is valuable information on every page."—*Mechanical World.*

"The information is precisely that likely to be required in practice. . . . The work forms a desirable addition to the library, not only of the works' manager, but of anyone connected with general engineering."—*Mining Journal.*

"A formidable mass of facts and figures, readily accessible through an elaborate index . . . . Such a volume will be found absolutely necessary as a book of reference in all sorts of 'works' connected with the metal trades. . . . Any ordinary foreman or workman can find all he wants in the crowded pages of this useful work."—*Ryland's Iron Trades Circular*

Engineering Construction.

PATTERN-MAKING: A Practical Treatise, embracing the Main Types of Engineering Construction, and including Gearing, both Hand and Machine made, Engine Work, Sheaves and Pulleys, Pipes and Columns, Screws, Machine Parts, Pumps and Cocks, the Moulding of Patterns in Loam and Greensand, &c., together with the methods of Estimating the weight of Castings; to which is added an Appendix of Tables for Workshop Reference. By a FOREMAN PATTERN MAKER. With upwards of Three Hundred and Seventy Illustrations. Crown 8vo, 7*s.* 6*d.* cloth.

"A well-written technical guide, evidently written by a man who understands and has practised what he has written about; he says what he has to say in a plain, straightforward manner. We cordially recommend the treatise to engineering students, young journeymen, and others desirous of being initiated into the mysteries of pattern-making."—*Builder.*

"We can confidently recommend this comprehensive treatise."—*Building News.*

"A valuable contribution to the literature of an important branch of engineering construction, which is likely to prove a welcome guide to many workmen, especially to draughtsmen who have lacked a training in the shops, pupils pursuing their practical studies in our factories, and to employers and managers in engineering works."—*Hardware Trade Journal.*

"More than 370 illustrations help to explain the text, which is, however, always clear and explicit, thus rendering the work an excellent *vade mecum* for the apprentice who desires to become master of his trade."—*English Mechanic.*

Smith's Tables for Mechanics, etc.

TABLES, MEMORANDA, AND CALCULATED RESULTS, FOR MECHANICS, ENGINEERS, ARCHITECTS, BUILDERS, etc. Selected and Arranged by FRANCIS SMITH. Third Edition, Revised and Enlarged, 250 pp., waistcoat-pocket size, 1*s.* 6*d.* limp leather.

"It would, perhaps, be as difficult to make a small pocket-book selection of notes and formulæ to suit ALL engineers as it would be to make a universal medicine; but Mr. Smith's waistcoat-pocket collection may be looked upon as a successful attempt."—*Engineer.*

"The best example we have ever seen of 250 pages of useful matter packed into the dimensions of a card-case."—*Building News.*

"A veritable pocket treasury of knowledge."—*Iron.*

The High-Pressure Steam Engine.

THE HIGH-PRESSURE STEAM-ENGINE: An Exposition of its Comparative Merits and an Essay towards an Improved System of Construction. By DR. ERNST ALBAN. Translated from the German, with Notes, by DR. POLE, M. Inst. C.E., &c. With 28 Plates. 8vo, 16*s.* 6*d.* cloth.

"Goes thoroughly into the examination of the high-pressure engine, the boiler, and its appendages, and deserves a place in every scientific library."—*Steam Shipping Chronicle.*

Steam Boilers.

A TREATISE ON STEAM BOILERS: Their Strength, Construction, and Economical Working. By ROBERT WILSON, C.E. Fifth Edition. 12mo, 6*s.* cloth.

"The best treatise that has ever been published on steam boilers."—*Engineer.*

"The author shows himself perfect master of his subject, and we heartily recommend all employing steam power to possess themselves of the work."—*Ryland's Iron Trade Circular.*

Boiler Making.

THE BOILER-MAKER'S READY RECKONER. With Examples of Practical Geometry and Templating, for the Use of Platers, Smiths and Riveters. By JOHN COURTNEY, Edited by D. K. CLARK, M.I.C.E. Second Edition, revised, with Additions, 12mo, 5*s.* half-bound.

"A most useful work. . . . . No workman or apprentice should be without this book."—*Iron Trade Circular.*

"A reliable guide to the working boiler-maker."—*Iron.*

"Boiler-makers will readily recognise the value of this volume. . . . The tables are clearly printed, and so arranged that they can be referred to with the greatest facility, so that it cannot be doubted that they will be generally appreciated and much used."—*Mining Journal.*

Steam Engine.

TEXT-BOOK ON THE STEAM ENGINE. By T. M. GOODEVE, M.A., Barrister-at-Law, Author of "The Elements of Mechanism," &c. Seventh Edition. With numerous Illustrations. Crown 8vo, 6*s.* cloth.

"Professor Goodeve has given us a treatise on the steam engine which will bear comparison with anything written by Huxley or Maxwell, and we can award it no higher praise."—*Engineer.*

Steam.

THE SAFE USE OF STEAM. Containing Rules for Unprofessional Steam-users. By an ENGINEER. Fifth Edition. Sewed, 6*d.*

"If steam-users would but learn this little book by heart, boiler explosions would become sensations by their rarity."—*English Mechanic.*

Coal and Speed Tables.

A POCKET BOOK OF COAL AND SPEED TABLES, for Engineers and Steam-users. By NELSON FOLEY, Author of "Boiler Construction." Pocket-size, 3*s.* 6*d.* cloth; 4*s.* leather.

"This is a very useful book, containing very useful tables. The results given are well chosen, and the volume contains evidence that the author really understands his subject. We can recommend the work with pleasure."—*Mechanical World.*

"These tables are designed to meet the requirements of every-day use; they are of sufficient scope for most practical purposes, and may be commended to engineers and users of steam."—*Iron.*

"This pocket-book well merits the attention of the practical engineer. Mr. Foley has compiled a very useful set of tables, the information contained in which is frequently required by engineers, coal consumers and users of steam."—*Iron and Coal Trades Review.*

Fire Engineering.

FIRES, FIRE-ENGINES, AND FIRE-BRIGADES. With a History of Fire-Engines, their Construction, Use, and Management; Remarks on Fire-Proof Buildings, and the Preservation of Life from Fire; Statistics of the Fire Appliances in English Towns; Foreign Fire Systems; Hints on Fire Brigades, &c. &c. By CHARLES F. T. YOUNG, C.E. With numerous Illustrations, 544 pp., demy 8vo, £1 4*s.* cloth.

"To such of our readers as are interested in the subject of fires and fire apparatus, we can most heartily commend this book. It is really the only English work we now have upon the subject."—*Engineering.*

"It displays much evidence of careful research; and Mr. Young has put his facts neatly ogether. It is evident enough that his acquaintance with the practical details of the construction of steam fire engines, old and new, and the conditions with which it is necessary hey should comply, ? accurate and full."—*Engineer.*

Gas Lighting.

COMMON SENSE FOR GAS-USERS: A Catechism of Gas-Lighting for Householders, Gasfitters, Millowners, Architects, Engineers, etc. By ROBERT WILSON, C.E., Author of "A Treatise on Steam Boilers." Second Edition. Crown 8vo, sewed, with Folding Plates and Wood Engravings, 2*s.* 6*d.*

"All gas-users will decidedly benefit, both in pocket and comfort, if they will avail themselves of Mr. Wilson's counsels."—*Engineering.*

THE POPULAR WORKS OF MICHAEL REYNOLDS

(*Known as* "THE ENGINE DRIVER'S FRIEND").

Locomotive-Engine Driving.

LOCOMOTIVE-ENGINE DRIVING: A Practical Manual for Engineers in charge of Locomotive Engines. By MICHAEL REYNOLDS, Member of the Society of Engineers, formerly Locomotive Inspector L. B. and S. C. R. Seventh Edition. Including a KEY TO THE LOCOMOTIVE ENGINE. With Illustrations and Portrait of Author. Crown 8vo, 4*s*. 6*d*. cloth.

"Mr. Reynolds has supplied a want, and has supplied it well. We can confidently recommend the book, not only to the practical driver, but to everyone who takes an interest in the performance of locomotive engines."—*The Engineer.*

"Were the cautions and rules given in the book to become part of the every-day working of our engine-drivers, we might have fewer distressing accidents to deplore."—*Scotsman.*

The Engineer, Fireman, and Engine-Boy.

THE MODEL LOCOMOTIVE ENGINEER, FIREMAN, and ENGINE-BOY. Comprising a Historical Notice of the Pioneer Locomotive Engines and their Inventors, with a project for the establishment of Certificates of Qualification in the Running Service of Railways. By MICHAEL REYNOLDS, Author of "Locomotive-Engine Driving." With numerous Illustrations and a fine Portrait of George Stephenson. Crown 8vo, 4*s*. 6*d*. cloth.

"From the technical knowledge of the author it will appeal to the railway man of to-day more forcibly than anything written by Dr. Smiles. . . . The volume contains information of a technical kind, and facts that every driver should be familiar with."—*English Mechanic.*

"We should be glad to see this book in the possession of everyone in the kingdom who has ever laid, or is to lay, hands on a locomotive engine."—*Iron.*

Stationary Engine Driving.

STATIONARY ENGINE DRIVING: A Practical Manual for Engineers in charge of Stationary Engines. By MICHAEL REYNOLDS. Third Edition, Enlarged. With Plates and Woodcuts. Crown 8vo, 4*s*. 6*d*. cloth.

"The author is thoroughly acquainted with his subjects, and his advice on the various points treated is clear and practical. . . . He has produced a manual which is an exceedingly useful one for the class for whom it is specially intended."—*Engineering.*

"Our author leaves no stone unturned. He is determined that his readers shall not only know something about the stationary engine, but all about it."—*Engineer.*

Continuous Railway Brakes.

CONTINUOUS RAILWAY BRAKES: A Practical Treatise on the several Systems in Use in the United Kingdom; their Construction and Performance. With copious Illustrations and numerous Tables. By MICHAEL REYNOLDS. Large crown 8vo, 9*s*. cloth.

"A popular explanation of the different brakes. It will be of great assistance in forming public opinion, and will be studied with benefit by those who take an interest in the brake."—*English Mechanic.*

"Written with sufficient technical detail to enable the p nc e and relative connection of the various parts of each particular brake to be readily grasped."—*Mechanical World.*

Engine-Driving Life.

ENGINE-DRIVING LIFE; or, Stirring Adventures and Incidents in the Lives of Locomotive-Engine Drivers. By MICHAEL REYNOLDS. Ninth Thousand. Crown 8vo, 2*s*. cloth.

"The book from first to last is perfectly fascinating. Wilkie Collins' most thrilling conceptions are thrown into the shade by true incidents, endless in their variety, related in every page."—*North British Mail.*

"Anyone who wishes to get a real insight into railway life cannot do better than read 'Engine-Driving Life' for himself; and if he once take it up he will find that the author's enthusiasm and real love of the engine-driving profession will carry him on till he has read every page."—*Saturday Review.*

Pocket Companion for Enginemen.

THE ENGINEMAN'S POCKET COMPANION AND PRACTICAL EDUCATOR FOR ENGINEMEN, BOILER ATTENDANTS, AND MECHANICS. By MICHAEL REYNOLDS, Mem. S. E., Author of "Locomotive Engine-Driving," "Stationary Engine-Driving," &c. With Forty-five Illustrations and numerous Diagrams. Royal 18mo, 3*s*. 6*d*., strongly bound in cloth for pocket wear. [*Just published.*

ARCHITECTURE, BUILDING, etc.

Construction.

THE SCIENCE OF BUILDING: An Elementary Treatise on the Principles of Construction. By E. WYNDHAM TARN, M.A., Architect. Second Edition, Revised, with 58 Engravings. Crown 8vo, 7s. 6d. cloth.

"A very valuable book, which we strongly recommend to all students."—*Builder.*

"No architectural student should be without this handbook of constructional knowledge."—*Architect.*

Villa Architecture.

A HANDY BOOK OF VILLA ARCHITECTURE: Being a Series of Designs for Villa Residences in various Styles. With Outline Specifications and Estimates. By C. WICKES, Architect, Author of "The Spires and Towers of England," &c. 30 Plates, 4to, half-morocco, gilt edges, £1 1s.

*** Also an Enlarged Edition of the above. 61 Plates, with Outline Specifications, Estimates, &c. £2 2s. half-morocco.

"The whole of the designs bear evidence of their being the work of an artistic architect, and they will prove very valuable and suggestive."—*Building News.*

Useful Text-Book for Architects.

THE ARCHITECT'S GUIDE: Being a Text-Book of Useful Information for Architects, Engineers, Surveyors, Contractors, Clerks of Works, &c. &c. By FREDERICK ROGERS, Architect, Author of "Specifications for Practical Architecture," &c. Second Edition, Revised and Enlarged. With numerous Illustrations. Crown 8vo, 6s. cloth.

"As a text-book of useful information for architects, engineers, surveyors, &c., it would be hard to find a handier or more complete little volume."—*Standard.*

"A young architect could hardly have a better guide-book."—*Timber Trades Journal.*

Taylor and Cresy's Rome.

THE ARCHITECTURAL ANTIQUITIES OF ROME. By the late G. L. TAYLOR, Esq., F.R.I.B.A., and EDWARD CRESY, Esq. New Edition, thoroughly revised by the Rev. ALEXANDER TAYLOR, M.A. (son of the late G. L. Taylor, Esq.), Fellow of Queen's College, Oxford, and Chaplain of Gray's Inn. Large folio, with 130 Plates, half-bound, £3 3s.

N.B.—This is the only book which gives on a large scale, and with the precision of architectural measurement, the principal Monuments of Ancient Rome in plan, elevation, and detail.

"Taylor and Cresy's work has from its first publication been ranked among those professional books which cannot be bettered. . . . It would be difficult to find examples of drawings, even among those of the most painstaking students of Gothic, more thoroughly worked out than are the one hundred and thirty plates in this volume."—*Architect.*

Drawing for Builders and Students in Architecture.

PRACTICAL RULES ON DRAWING, for the Operative Builder and Young Student in Architecture. By GEORGE PYNE. With 14 Plates, 4to, 7s. 6d. boards.

Civil Architecture.

THE DECORATIVE PART OF CIVIL ARCHITECTURE. By Sir WILLIAM CHAMBERS, F.R.S. With Illustrations, Notes, and an Examination of Grecian Architecture, by JOSEPH GWILT, F.S.A. Edited by W. H. LEEDS. 66 Plates, 4to, 21s. cloth.

The House-Owner's Estimator.

THE HOUSE-OWNER'S ESTIMATOR; or, What will it Cost to Build, Alter, or Repair? A Price Book adapted to the Use of Unprofessional People, as well as for the Architectural Surveyor and Builder. By the late JAMES D. SIMON, A.R.I.B.A. Edited and Revised by FRANCIS T. W. MILLER, A.R.I.B.A. With numerous Illustrations. Third Edition, Revised. Crown 8vo, 3s. 6d. cloth.

"In two years it will repay its cost a hundred times over."—*Field.*

"A very handy book."—*English Mechanic.*

Designing, Measuring, and Valuing.

THE STUDENT'S GUIDE to the PRACTICE of MEASURING AND VALUING ARTIFICERS' WORKS. Containing Directions for taking Dimensions, Abstracting the same, and bringing the Quantities into Bill, with Tables of Constants, and copious Memoranda for the Valuation of Labour and Materials in the respective Trades of Bricklayer and Slater, Carpenter and Joiner, Painter and Glazier, Paperhanger, &c. With 8 Plates and 63 Woodcuts. Originally edited by EDWARD DOBSON, Architect. Fifth Edition, Revised, with considerable Additions on Mensuration and Construction, and a New Chapter on Dilapidations, Repairs, and Contracts, by E. WYNDHAM TARN, M.A. Crown 8vo, 9s. cloth.

"Well fulfils the promise of its title-page, and we can thoroughly recommend it to the class for whose use it has been compiled. Mr. Tarn's additions and revisions have much increased the usefulness of the work, and have especially augmented its value to students."—*Engineering.*

"The work has been carefully revised and edited by Mr. E. Wyndham Tarn, M.A., and comprises several valuable additions on construction, mensuration, dilapidations and repairs, and other matters. . . . This edition will be found the most complete treatise on the principles of measuring and valuing artificers' work that has yet been published."—*Building News.*

Pocket Estimator.

THE POCKET ESTIMATOR for the BUILDING TRADES. Being an Easy Method of Estimating the various parts of a Building collectively, more especially applied to Carpenters' and Joiners' work. By A. C. BEATON, Author of "Quantities and Measurements." Third Edition, carefully revised, 33 Woodcuts, leather, waistcoat-pocket size, 1s. 6*d.*

"Contains a good deal of information not easily to be obtained from the ordinary price books. The prices given are accurate, and up to date."—*Building News.*

Builder's and Surveyor's Pocket Technical Guide.

THE POCKET TECHNICAL GUIDE AND MEASURER FOR BUILDERS AND SURVEYORS. Containing a Complete Explanation of the Terms used in Building Construction, Memoranda for Reference, Technical Directions for Measuring Work in all the Building Trades, with a Treatise on the Measurement of Timber, Complete Specifications, &c. &c. By A. C. BEATON. Second Edition, with 19 Woodcuts, leather, waistcoat-pocket size, 1*s.* 6*d.*

"An exceedingly handy pocket companion, thoroughly reliable."—*Builder's Weekly Reporter.*

"This neat little compendium contains all that is requisite in carrying out contracts for excavating, tiling, bricklaying, paving, &c."—*British Trade Journal.*

Donaldson on Specifications.

THE HANDBOOK OF SPECIFICATIONS; or, Practical Guide to the Architect, Engineer, Surveyor, and Builder, in drawing up Specifications and Contracts for Works and Constructions. Illustrated by Precedents of Buildings actually executed by eminent Architects and Engineers. By Professor T. L. DONALDSON, P.R.I.B.A., &c. New Edition, in One large Vol., 8vo, with upwards of 1,000 pages of Text, and 33 Plates, £1 11*s.* 6*d.* cloth.

"In this work forty-four specifications of executed works are given, including the specifications for parts of the new Houses of Parliament, by Sir Charles Barry, and for the new Royal Exchange, by Mr. Tite, M.P. The latter, in particular, is a very complete and remarkable document. It embodies, to a great extent, as Mr. Donaldson mentions, 'the bill of quantities with the description of the works.' . . . It is valuable as a record, and more valuable still as a book of precedents. . . . Suffice it to say that Donaldson's 'Handbook of Specifications' must be bought by all architects."—*Builder.*

Bartholomew and Rogers' Specifications.

SPECIFICATIONS FOR PRACTICAL ARCHITECTURE: A Guide to the Architect, Engineer, Surveyor, and Builder; with an Essay on the Structure and Science of Modern Buildings. Upon the Basis of the Work by ALFRED BARTHOLOMEW, thoroughly Revised, Corrected, and greatly added to by FREDERICK ROGERS, Architect. Second Edition, Revised, with Additions. With numerous Illusts., medium 8vo, 15*s.* cloth. [*Just published.*

"The collection of specifications prepared by Mr. Rogers on the basis of Bartholomew's work is too well known to need any recommendation from us. It is one of the books with which every young architect must be equipped; for time has shown that the specifications cannot be set aside through any defect in them."—*Architect.*

"Good forms for specifications are of considerable value, and it was an excellent idea to compile a work on the subject upon the basis of the late Alfred Bartholomew's valuable work. The second edition of Mr. Rogers's book is evidence of the want of a book dealing with modern requirements and materials."—*Building News.*

DECORATIVE ARTS, etc.

Woods and Marbles (Imitation of).

SCHOOL OF PAINTING FOR THE IMITATION OF WOODS AND MARBLES, as Taught and Practised by A. R. VAN DER BURG and P. VAN DER BURG, Directors of the Rotterdam Painting Institution. Second and Cheaper Edition. Royal folio, 18½ by 12½ in., Illustrated with 24 full-size Coloured Plates; also 12 plain Plates, comprising 154 Figures, price £1 11s. 6*d*.

List of Contents

Introductory Chapter — Tools required for Wood Painting—Observations on the different species of Wood: Walnut—Observations on Marble in general—Tools required for Marble Painting—St. Remi Marble: Preparation of the Paints: Process of Working—Wood Graining: Preparation of Stiff and Flat Brushes: Sketching different Grains and Knots: Glazing of Wood—Ash: Painting of Ash—Breche (Breccia) Marble: Breche Violette: Process of Working—Maple: Process of Working—The different species of White Marble: Methods of Working: Painting White Marble with Lac-dye: Painting White Marble with Poppy-paint—Mahogany: Methods of Working—Yellow Sienna Marble: Process of Working—Juniper: Characteristics of the Natural Wood: Method of Imitation—Vert de Mer Marble: Description of the Marble: Process of Working—Oak: Description of the varieties of Oak: Manipulation of Oak-painting: Tools employed: Method of Working—Waulsort Marble: Varieties of the Marble: Process of Working—The Painting of Iron with Red Lead: How to make Putty: Out-door Work: Varnishing: Priming and Varnishing Woods and Marbles: Painting in General: Ceilings and Walls: Gilding: Transparencies, Flags, &c.

List of Plates.

1. Various Tools required for Wood Painting—2, 3. Walnut: Preliminary Stages of Graining and Finished Specimen—4. Tools used for Marble Painting and Method of Manipulation—5, 6. St. Remi Marble: Earlier Operations and Finished Specimen—7. Methods of Sketching different Grains, Knots, &c.—8, 9. Ash: Preliminary Stages and Finished Specimen—10. Methods of Sketching Marble Grains—11, 12. Breche Marble: Preliminary Stages of Working and Finished Specimen—13. Maple: Methods of Producing the different Grains—14, 15. Bird's-eye Maple: Preliminary Stages and Finished Specimen—16. Methods of Sketching the different Species of White Marble—17, 18. White Marble: Preliminary Stages of Process and Finished Specimen—19. Mahogany: Specimens of various Grains and Methods of Manipulation—20, 21. Mahogany: Earlier Stages and Finished Specimen—22, 23, 24. Sienna Marble: Varieties of Grain, Preliminary Stages and Finished Specimen—25, 26, 27. Juniper Wood: Methods of producing Grain, &c.: Preliminary Stages and Finished Specimen—28, 29, 30. Vert de Mer Marble: Varieties of Grain and Methods of Working Unfinished and Finished Specimens—31, 32, 33. Oak: Varieties of Grain, Tools Employed, and Methods of Manipulation, Preliminary Stages and Finished Specimen—34, 35, 36. Waulsort Marble: Varieties of Grain, Unfinished and Finished Specimens.

"Those who desire to attain skill in the art of painring woods and marbles, will find advantage in consulting this book. . . . Some of the Working Men's Clubs should give their young men the opportunity to study it."—*Builder*.

"A comprehensive guide to the art. The explanations of the processes, the manipulation and management of the colours, and the beautifully executed plates will not be the least valuable to the student who aims at making his work a faithful transcript of nature."—*Building News*.

Colour.

A GRAMMAR OF COLOURING. Applied to Decorative Painting and the Arts. By GEORGE FIELD. New Edition, adapted to the use of the Ornamental Painter and Designer. By ELLIS A. DAVIDSON. With New Coloured Diagrams and Engravings. 12mo, 3*s*. 6*d*. cloth boards.

"The book is a most useful *resumé* of the properties of pigments."—*Builder*.

House Decoration.

ELEMENTARY DECORATION. A Guide to the Simpler Forms of Everyday Art, as applied to the Interior and Exterior Decoration of Dwelling Houses, &c. By JAMES W. FACEY. With 68 Cuts. 2*s*. cloth limp.

"As a technical guide-book to the decorative painter it will be found reliable."—*Building News*.

*** *By the same Author, just published.*

PRACTICAL HOUSE DECORATION: A Guide to the Art of Ornamental Painting, the Arrangement of Colours in Apartments, and the principles of Decorative Design. With some Remarks upon the Nature and Properties of Pigments. With numerous Illustrations. 12mo, 2*s*. 6*d*. cl. limp

N.B.—*The above Two Works together in One Vol., strongly half-bound*, 5*s*.

House Painting, etc.

HOUSE PAINTING, GRAINING, MARBLING, AND SIGN WRITING, A Practical Manual of. By ELLIS A. DAVIDSON. Fourth Edition. With Coloured Plates and Wood Engravings. 12mo, 6*s*. cloth boards.

A mass of information, of use to the amateur and of value to the practical man."—*English Mechanic*.

DELAMOTTES' WORKS on ILLUMINATION & ALPHABETS.

A PRIMER OF THE ART OF ILLUMINATION, for the Use of Beginners: with a Rudimentary Treatise on the Art, Practical Directions for its exercise, and Examples taken from Illuminated MSS., printed in Gold and Colours. By F. DELAMOTTE. New and cheaper edition. Small 4to, 6s. ornamental boards.

". . . . The examples of ancient MSS. recommended to the student, which, with much good sense, the author chooses from collections accessible to all, are selected with judgment and knowledge, as well as taste."—*Athenæum.*

ORNAMENTAL ALPHABETS, Ancient and Mediæval, from the Eighth Century, with Numerals; including Gothic, Church-Text, large and small, German, Italian, Arabesque, Initials for Illumination, Monograms Crosses, &c. &c., for the use of Architectural and Engineering Draughtsmen, Missal Painters, Masons, Decorative Painters, Lithographers, Engravers, Carvers, &c. &c. Collected and Engraved by F. DELAMOTTE, and printed in Colours. New and Cheaper Edition. Royal 8vo, oblong, 2s. 6*d.* ornamental boards.

'For those who insert enamelled sentences round gilded chalices, who blazon shop legends over shop-doors, who letter church walls with pithy sentences from the Decalogue, this book will be useful."—*Athenæum.*

EXAMPLES OF MODERN ALPHABETS, Plain and Ornamental; including German, Old English, Saxon, Italic, Perspective, Greek, Hebrew, Court Hand, Engrossing, Tuscan, Riband, Gothic, Rustic, and Arabesque; with several Original Designs, and an Analysis of the Roman and Old English Alphabets, large and small, and Numerals, for the use of Draughtsmen, Surveyors, Masons, Decorative Painters, Lithographers, Engravers, Carvers, &c. Collected and Engraved by F. DELAMOTTE, and printed in Colours. New and Cheaper Edition. Royal 8vo, oblong, 2s. 6*d.* ornamental boards.

"There is comprised in it every possible shape into which the letters of the alphabet and numerals can be formed, and the talent which has been expended in the conception of the various plain and ornamental letters is wonderful."—*Standard.*

MEDIÆVAL ALPHABETS AND INITIALS FOR ILLUMINATORS. By F. G. DELAMOTTE. Containing 21 Plates and Illuminated Title, printed in Gold and Colours. With an Introduction by J. WILLIS BROOKS. Fourth and cheaper edition. Small 4to, 4s. ornamental boards.

"A volume in which the letters of the alphabet come forth glorified in gilding and all the colours of the prism interwoven and intertwined and intermingled."—*Sun.*

THE EMBROIDERER'S BOOK OF DESIGN. Containing Initials, Emblems, Cyphers, Monograms, Ornamental Borders, Ecclesiastical Devices, Mediæval and Modern Alphabets, and National Emblems. Collected by F. DELAMOTTE, and printed in Colours. Oblong royal 8vo, 1s. 6*d.*, ornamental wrapper.

"The book will be of great assistance to ladies and young children who are endowed with the art of plying the needle in this most ornamental and useful pretty work."—*East Anglian Times.*

Wood Carving.

INSTRUCTIONS IN WOOD-CARVING, for Amateurs; with Hints on Design. By A LADY. With Ten large Plates, 2s. 6*d.* in emblematic wrapper.

"The handicraft of the wood-carver, so well as a book can impart it, may be learnt from 'A Lady's' publication."—*Athenæum.*

"The directions given are plain and easily understood."—*English Mechanic.*

Glass Painting.

GLASS STAINING AND THE ART OF PAINTING ON GLASS. From the German of Dr. GESSERT and EMANUEL OTTO FROMBERG. With an Appendix on THE ART OF ENAMELLING. 12mo, 2s. 6*d.* cloth limp.

Letter Painting.

THE ART OF LETTER PAINTING MADE EASY. By JAMES GREIG BADENOCH. With 12 full-page Engravings of Examples, 1s. cloth limp.

"The system is a simple one, but quite original, and well worth the careful attention of letter-painters. It can be easily mastered and remembered."—*Building News.*

CARPENTRY, TIMBER, etc.

Tredgold's Carpentry, partly Re-written and Enlarged by Tarn.

THE ELEMENTARY PRINCIPLES OF CARPENTRY. A Treatise on the Pressure and Equilibrium of Timber Framing, the Resistance of Timber, and the Construction of Floors, Arches, Bridges, Roofs, Uniting Iron and Stone with Timber, &c. To which is added an Essay on the Nature and Properties of Timber, &c., with Descriptions of the kinds of Wood used in Building; also numerous Tables of the Scantlings of Timber for different purposes, the Specific Gravities of Materials, &c. By THOMAS TREDGOLD, C.E. With an Appendix of Specimens of Various Roofs of Iron and Stone, Illustrated. Seventh Edition, thoroughly revised and considerably enlarged by E. WYNDHAM TARN, M.A., Author of "The Science of Building," &c. With 61 Plates, Portrait of the Author, and several Woodcuts. In one large vol., 4to, price £1 5s. cloth. [*Just published.*

"Ought to be in every architect's and every builder's library."—*Builder.*

"A work whose monumental excellence must commend it wherever skilful carpentry is concerned. The author's principles are rather confirmed than impaired by time. The additional plates are of great intrinsic value."—*Building News.*

Woodworking Machinery.

WOODWORKING MACHINERY: Its Rise, Progress, and Construction. With Hints on the Management of Saw Mills and the Economical Conversion of Timber. Illustrated with Examples of Recent Designs by leading English, French, and American Engineers. By M. POWIS BALE, A.M. Inst. C.E., M.I.M.E. Large crown 8vo, 12s. 6d. cloth.

"Mr. Bale is evidently an expert on the subject, and he has collected so much information that his book is all-sufficient for builders and others engaged in the conversion of timber."—*Architect.*

"The most comprehensive compendium of wood-working machinery we have seen. The author is a thorough master of his subject."—*Building News.*

"The appearance of this book at the present time will, we should think, give a considerable impetus to the onward march of the machinist engaged in the designing and manufacture of wood-working machines. It should be in the office of every wood-working factory."—*English Mechanic.*

Saw Mills.

SAW MILLS: Their Arrangement and Management, and the Economical Conversion of Timber. (Being a Companion Volume to "Woodworking Machinery.") By M. POWIS BALE, A.M. Inst. C.E., M.I.M.E. With numerous Illustrations. Crown 8vo, 10s. 6d. cloth.

"The author is favourably known by his former work on 'Woodworking Machinery,' of which we were able to speak approvingly. This is a companion volume, in which the *administration* of a large sawing establishment is discussed, and the subject examined from a financial standpoint Hence the size, shape, order, and disposition of saw-mills and the like are gone into in detail, and the course of the timber is traced from its reception to its delivery in its converted state. We could not desire a more complete or practical treatise."—*Builder.*

"We highly recommend Mr. Bale's work to the attention and perusal of all those who are engaged in the art of wood conversion, or who are about building or remodelling saw-mills on improved principles."—*Building News.*

Carpentering.

THE CARPENTER'S NEW GUIDE; or, Book of Lines for Carpenters; comprising all the Elementary Principles essential for acquiring a knowledge of Carpentry. Founded on the late PETER NICHOLSON'S Standard Work. A New Edition, revised by ARTHUR ASHPITEL, F.S.A. Together with Practical Rules on Drawing, by GEORGE PYNE. With 74 Plates, 4to, £1 1s. cloth.

Handrailing.

A PRACTICAL TREATISE ON HANDRAILING: Showing New and Simple Methods for Finding the Pitch of the Plank, Drawing the Moulds, Bevelling, Jointing-up, and Squaring the Wreath. By GEORGE COLLINGS. Illustrated with Plates and Diagrams. 12mo, 1s. 6d. cloth limp.

Circular Work.

CIRCULAR WORK IN CARPENTRY AND JOINERY: A Practical Treatise on Circular Work of Single and Double Curvature. By GEORGE COLLINGS, Author of "A Practical Treatise on Handrailing." Illustrated with numerous Diagrams. 12mo, 2s. 6d. cloth limp. [*Just published.*

Timber Merchant's Companion.

THE TIMBER MERCHANT'S AND BUILDER'S COMPANION. Containing New and Copious Tables of the Reduced Weight and Measurement of Deals and Battens, of all sizes, from One to a Thousand Pieces, and the relative Price that each size bears per Lineal Foot to any given Price per Petersburg Standard Hundred; the Price per Cube Foot of Square Timber to any given Price per Load of 50 Feet; the proportionate Value of Deals and Battens by the Standard, to Square Timber by the Load of 50 Feet; the readiest mode of ascertaining the Price of Scantling per Lineal Foot of any size, to any given Figure per Cube Foot, &c. &c. By WILLIAM DOWSING. Third Edition, Revised and Corrected. Crown 8vo, 3*s.* cloth.

"Everything is as concise and clear as it can possibly be made. There can be no doubt that every timber merchant and builder ought to possess it."—*Hull Advertiser.*

"An exceedingly well-arranged, clear, and concise manual of tables for the use of all who buy or sell timber."—*Journal of Forestry.*

Practical Timber Merchant.

THE PRACTICAL TIMBER MERCHANT. Being a Guide for the use of Building Contractors, Surveyors, Builders, &c., comprising useful Tables for all purposes connected with the Timber Trade, Marks of Wood, Essay on the Strength of Timber, Remarks on the Growth of Timber, &c. By W. RICHARDSON. Fcap. 8vo, 3*s.* 6*d.* cloth.

"This handy manual contains much valuable information for the use of timber merchants, builders, foresters, and all others connected with the growth, sale, and manufacture of timber."—*Journal of Forestry.*

Timber Freight Book.

THE TIMBER MERCHANT'S, SAW MILLER'S, AND IMPORTER'S FREIGHT BOOK AND ASSISTANT. Comprising Rules, Tables, and Memoranda relating to the Timber Trade. By WILLIAM RICHARDSON Timber Broker; together with a Chapter on "Speeds of Saw Mill Machinery," by M. POWIS BALE, M.I.M.E., &c. 12mo, 3*s.* 6*d.* cloth boards.

"A very useful manual of rules, tables, and memoranda, relating to the timber trade. We recommend it as a compendium of calculation to all timber measurers and merchants, and as supplying a real want in the trade."—*Building News.*

Tables for Packing-Case Makers.

PACKING-CASE TABLES; showing the number of Superficial Feet in Boxes or Packing-Cases, from six inches square and upwards. By W. RICHARDSON, Timber Broker. Oblong 4to, 3*s.* 6*d.* cloth.

"Invaluable labour-saving tables."—*Ironmonger*

"Will save much labour and calculation."—*Grocer.*

Superficial Measurement.

THE TRADESMAN'S GUIDE TO SUPERFICIAL MEASUREMENT. Tables calculated from 1 to 200 inches in length, by 1 to 108 inches in breadth. For the use of Architects, Engineers, Timber Merchants, Builders, &c. By JAMES HAWKINGS. Third Edition. Fcap., 3*s.* 6*d.* cloth.

"A useful collection of tables to facilitate rapid calculation of surfaces. The exact area of any surface of which the limits have been ascertained can be instantly determined. The book will be found of the greatest utility to all engaged in building operations."—*Scotsman.*

Forestry.

THE ELEMENTS OF FORESTRY. Designed to afford Information concerning the Planting and Care of Forest Trees for Ornament or Profit, with Suggestions upon the Creation and Care of Woodlands. By F. B. HOUGH. Large crown 8vo, 10*s.* cloth.

Timber Importer's Guide.

THE TIMBER IMPORTER'S, TIMBER MERCHANT'S AND BUILDER'S STANDARD GUIDE. By RICHARD E. GRANDY. Comprising an Analysis of Deal Standards, Home and Foreign, with Comparative Values and Tabular Arrangements for fixing Nett Landed Cost on Baltic and North American Deals, including all intermediate Expenses, Freight, Insurance, &c. &c. Second Edition, carefully revised. 12mo, 3*s.* 6*d.* cloth.

"Everything it pretends to be: built up gradually, it leads one from a forest to a treenail, and throws in, as a makeweight, a host of material concerning bricks, columns, cisterns, &c."—*English Mechanic.*

MINING AND MINING INDUSTRIES.

Metalliferous Mining.

BRITISH MINING: A Treatise on the History, Discovery, Practical Development, and Future Prospects of Metalliferous Mines in the United Kingdom. By ROBERT HUNT, F.R.S., Keeper of Mining Records; Editor of "*Ure's* Dictionary of Arts, Manufactures, and Mines," &c. Upwards of 950 pp., with 230 Illustrations. Super-royal 8vo, £3 3s. cloth.

*** OPINIONS OF THE PRESS.

"One of the most valuable works of reference of modern times. Mr. Hunt, as keeper of mining records of the United Kingdom, has had opportunities for such a task not enjoyed by anyone else, and has evidently made the most of them. . . . The language and style adopted are good, and the treatment of the various subjects laborious, conscientious, and scientific."—*Engineering.*

"Probably no one in this country was better qualified than Mr. Hunt for undertaking such a work. Brought into frequent and close association during a long life-time with the principal guardians of our mineral and metallurgical industries, he enjoyed a position exceptionally favourable for collecting the necessary information. The use which he has made of his opportunities is sufficiently attested by the dense mass of information crowded into the handsome volume which has just been published. . . . In placing before the reader a sketch of the present position of British Mining, Mr. Hunt treats his subject so fully and illustrates it so amply that this section really forms a little treatise on practical mining. . . . The book is, in fact, a treasure-house of statistical information on mining subjects, and we know of no other work embodying so great a mass of matter of this kind. Were this the only merit of Mr. Hunt's volume it would be sufficient to render it indispensable in the library of everyone interested in the development of the mining and metallurgical industries of this country."—*Athenæum.*

"A mass of information not elsewhere available, and of the greatest value to those who may be interested in our great mineral industries."—*Engineer.*

"A sound, business-like collection of interesting facts. . . . The amount of information Mr. Hunt has brought together is enormous. . . . The volume appears likely to convey more instruction upon the subject than any work hitherto published."—*Mining Journal.*

"The work will be for the mining industry what Dr. Percy's celebrated treatise has been for the metallurgical—a book that cannot with advantage be omitted from the library."—*Iron and Coal Trades' Review.*

"The literature of mining has hitherto possessed no work approaching in importance to that which has just been published. There is much in Mr. Hunt's valuable work that every shareholder in a mine should read with close attention. The entire subject of practical mining—from the first search for the lode to the latest stages of dressing the ore—is dealt with in a masterly manner." —*Academy.*

Coal and Iron.

THE COAL AND IRON INDUSTRIES OF THE UNITED KINGDOM. Comprising a Description of the Coal Fields, and of the Principal Seams of Coal, with Returns of their Produce and its Distribution, and Analyses of Special Varieties. Also an Account of the occurrence of Iron Ores in Veins or Seams; Analyses of each Variety; and a History of the Rise and Progress of Pig Iron Manufacture since the year 1740, exhibiting the Economies introduced in the Blast Furnaces for its Production and Improvement. By RICHARD MEADE, Assistant Keeper of Mining Records. With Maps of the Coal Fields and Ironstone Deposits of the United Kingdom. 8vo, £1 8s. cloth.

"The book is one which must find a place on the shelves of all interested in coal and iron production, and in the iron, steel, and other metallurgical industries."—*Engineer.*

"Of this book we may unreservedly say that it is the best of its class which we have ever met. . . . A book of reference which no one engaged in the iron or coal trades should omit from his library."—*Iron and Coal Trades' Review.*

"An exhaustive treatise and a valuable work f reference."—*Mining Journal.*

Prospecting.

THE PROSPECTOR'S HANDBOOK: A Guide for the Prospector and Traveller in Search of Metal-Bearing or other Valuable Minerals. By J. W. ANDERSON, M.A. (Camb.), F.R.G.S., Author of "Fiji and New Caledonia." Small crown 8vo, 3s. 6d. cloth. [*Just published.*

"Will supply a much felt want, especially among Colonists, in whose way are so often thrown many mineralogical specimens, the value of which it is difficult for anyone, not a specialist, to determine. The author has placed his instructions before his readers in the plainest possible terms, and his book is the best of its kind."—*Engineer.*

"How to find commercial minerals, and how to identify them they are found, are the leading points to which attention is directed. The author has manag d to pack as much practical detail into his pages as would supply material for a book three times its size."—*Mining Journal.*

"Those toilers who explore the trodden or untrodden tracks on the face of the globe will find much that is useful to them in this book."—*Athenæum.*

Metalliferous Minerals and Mining.

TREATISE ON METALLIFEROUS MINERALS AND MINING. By D. C. Davies, F.G.S., Mining Engineer, &c., Author of "A Treatise on Slate and Slate Quarrying." Illustrated with numerous Wood Engravings. Second Edition, carefully Revised. Crown 8vo, 12*s.* 6*d.* cloth.

"Neither the practical miner nor the general reader interested in mines, can have a better book for his companion and his guide."—*Mining Journal.*

"The volume is one which no student of mineralogy should be without."—*Colliery Guardian.*

"We are doing our readers a service in calling their attention to this valuable work."—*Mining World.*

"A book that will not only be useful to the geologist, the practical miner, and the metallurgist: but also very interesting to the general public."—*Iron.*

"As a history of the present state of mining throughout the world this book has a real value, and it supplies an actual want, for no such information has hitherto been brought together within such limited space."—*Athenæum.*

Earthy Minerals and Mining.

A TREATISE ON EARTHY AND OTHER MINERALS AND MINING. By D. C. Davies, F.G.S. Uniform with, and forming a Companion Volume to, the same Author's "Metalliferous Minerals and Mining." With 76 Wood Engravings. Crown 8vo, 12*s.* 6*d.* cloth.

"It is essentially a practical work, intended primarily for the use of practical men. . . . We do not remember to have met with any English work on mining matters that contains the same amount of information packed in equally convenient form."—*Academy.*

'The book is clearly the result of many years' careful work and thought, and we should be inclined to rank it as among the very best of the handy technical and trades manuals which have recently appeared."—*British Quarterly Review.*

"The subject matter of the volume will be found of high value by all—and they are a numerous class—who trade in earthy minerals."—*Athenæum.*

"Will be found of permanent value for information and reference."—*Iron.*

Underground Pumping Machinery.

MINE DRAINAGE. Being a Complete and Practical Treatise on Direct-Acting Underground Steam Pumping Machinery, with a Description of a large number of the best known Engines, their General Utility and the Special Sphere of their Action, the Mode of their Application, and their merits compared with other forms of Pumping Machinery. By Stephen Michell. 8vo, 15*s.* cloth.

"Will be highly esteemed by colliery owners and lessees, mining engineers, and students generally who require to be acquainted with the best means of securing the drainage of mines. It is a most valuable work, and stands almost alone in the literature of steam pumping machinery.'—*Colliery Guardian.*

"Much valuable information is given, so that the book is thoroughly worthy of an extensive circulation amongst practical men and purchasers of machinery."—*Mining Journal.*

Mining Tools.

A MANUAL OF MINING TOOLS. For the Use of Mine Managers, Agents, Students, &c. By William Morgans, Lecturer on Practical Mining at the Bristol School of Mines. 12mo, 3*s.* cloth boards.

ATLAS OF ENGRAVINGS to Illustrate the above, containing 235 Illustrations of Mining Tools, drawn to scale. 4to, 6*s.* cloth boards.

"Students in the science of mining, and overmen, captains, managers, and viewers may gain practical knowledge and useful hints by the study of Mr. Morgans' manual."—*Colliery Guardian.*

"A valuable work, which will tend materially to improve our mining literature."—*Mining Journal.*

Coal Mining.

COAL AND COAL MINING: A Rudimentary Treatise on. By Warington W. Smyth, M.A., F.R.S., &c., Chief Inspector of the Mines of the Crown. New Edition, Revised and Corrected. With numerous Illustrations. 12mo, 4*s.* cloth boards.

"As an outline is given of every known coal-field in this and other countries, as well as of the principal methods of working, the book will doubtless interest a very large number of readers."—*Mining Journal.*

Subterraneous Surveying.

SUBTERRANEOUS SURVEYING, Elementary and Practical Treatise on; with and without the Magnetic Needle. By Thomas Fenwick, Surveyor of Mines, and Thomas Baker, C.E 12mo, 3*s.* cloth boards.

NAVAL ARCHITECTURE, NAVIGATION, etc.

Chain Cables.

CHAIN CABLES AND CHAINS. Comprising Sizes and Curves of Links, Studs, &c., Iron for Cables and Chains, Chain Cable and Chain Making, Forming and Welding Links, Strength of Cables and Chains, Certificates for Cables, Marking Cables, Prices of Chain Cables and Chains, Historical Notes, Acts of Parliament, Statutory Tests, Charges for Testing, List of Manufacturers of Cables, &c., &c. By THOMAS W. TRAILL, F.E.R.N., M. Inst. C.E., the Engineer Surveyor in Chief, Board of Trade, the Inspector of Chain Cable and Anchor Proving Establishments, and General Superintendent, Lloyd's Committee on Proving Establishments. With numerous Tables, Illustrations and Lithographic Drawings. Folio, £2 2s. cloth, bevelled boards.

"The author writes not only with a full acquaintance with scientific formulæ and details, but also with a profound and fully-instructed sense of the importance to the safety of our ships and sailors of fidelity in the manufacture of cables. We heartily recommend the book to the specialists to whom it is addressed."—*Athenæum.*

"It contains a vast amount of valuable information. Nothing seems to be wanting to make a complete and standard work of reference on the subject."—*Nautical Magazine.*

Pocket-Book for Naval Architects and Shipbuilders.

THE NAVAL ARCHITECT'S AND SHIPBUILDER'S POCKET-BOOK of Formulæ, Rules, and Tables, and Marine Engineer's and Surveyor's Handy Book of Reference. By CLEMENT MACKROW, Member of. the Institution of Naval Architects, Naval Draughtsman. Third Edition, Revised. With numerous Diagrams, &c. Fcap., 12s. 6d. strongly bound in leather.

"Should be used by all who are engaged in the construction or design of vessels. . . . Will be found to contain the most useful tables and formulæ required by shipbuilders, carefully collected from the best authorities, and put together in a popular and simple form."—*Engineer.*

"The professional shipbuilder has now, in a convenient and accessible form, reliable data for solving many of the numerous problems that present themselves in the course of his work."—*Iron.*

"There is scarcely a subject on which a naval architect or shipbuilder can require to refresh his memory which will not be found within the covers of Mr. Mackrow's book."—*English Mechanic.*

Pocket-Book for Marine Engineers.

A POCKET-BOOK OF USEFUL TABLES AND FORMULÆ FOR MARINE ENGINEERS. By FRANK PROCTOR A.I.N.A. Third Edition. Royal 32mo, leather, gilt edges, with strap, 4s.

"We recommend it to our readers as going far to supply a long-felt want."—*Naval Science.*
"A most useful companion to all marine engineers."—*United Service Gazette.*

Lighthouses.

EUROPEAN LIGHTHOUSE SYSTEMS. Being a Report of a Tour of Inspection made in 1873. By Major GEORGE H. ELLIOT, Corps of Engineers, U.S.A. With 51 Engravings and 31 Woodcuts. 8vo, 21s. cloth.

*** *The following are published in* WEALE'S RUDIMENTARY SERIES.

MASTING, MAST-MAKING, AND RIGGING OF SHIPS. By ROBERT KIPPING, N.A. Fifteenth Edition. 12mo, 2s. 6d. cloth boards.

SAILS AND SAIL-MAKING. Eleventh Edition, Enlarged, with an Appendix. By ROBERT KIPPING, N.A. Illustrated. 12mo, 3s. cloth boards.

NAVAL ARCHITECTURE. By JAMES PEAKE. Fifth Edition with Plates and Diagrams. 12mo, 4s. cloth boards.

MARINE ENGINES AND STEAM VESSELS (A Treatise on). By ROBERT MURRAY, C.E., Principal Officer to the Board of Trade for the East Coast of Scotland District. Eighth Edition, thoroughly Revised, with considerable Additions, by the Author and by GEORGE CARLISLE, C.E., Senior Surveyor to the Board of Trade at Liverpool. 12mo, 5s. cloth boards.

PRACTICAL NAVIGATION. Consisting of the Sailor's Sea-Book, by JAMES GREENWOOD and W. H. ROSSER; together with the requisite Mathematical and Nautical Tables for the Working of the Problems, by HENRY LAW, C.E., and Professor J. R. YOUNG. 12mo, 7s., half-bound

NATURAL PHILOSOPHY AND SCIENCE.

Text Book of Electricity.

THE STUDENT'S TEXT-BOOK OF ELECTRICITY. HENRY M. NOAD, Ph.D., F.R.S., F.C.S. New Edition, carefully Rev With an Introduction and Additional Chapters, by W. H. PREECE, M.I.C Vice-President of the Society of Telegraph Engineers, &c. With 470 Illu tions. Crown 8vo, 12*s.* 6*d.* cloth.

"The original plan of this book has been carefully adhered to so as to make it a refle existing state of electrical science, adapted for students. . . . Discovery seems to ha gressed with marvellous strides; nevertheless it has now apparently ceased, and practical tions have commenced their career; and it is to give a faithful account of these that th edition of Dr. Noad's valuable text-book is launched forth."—*Extract from Introduction by Preece, Esq.*

"We can recommend Dr. Noad's book for clear style, great range of subject, a good and a plethora of woodcuts. Such collections as the present are indispensable."—*Athenæ*

"An admirable text-book for every student—beginner or advanced—of electri *Engineering.*

"Dr. Noad's text-book has earned for itself the reputation of a truly scientific manual student of electricity, and we gladly hail this new amended edition, which brings it once the front. Mr. Preece as reviser, with the assistance of Mr. H. R. Kempe and Mr. J. P. E has added all the practical results of recent invention and research to the admirable th expositions of the author, so that the book is about as complete and advanced as it is pos any book to be within the limits of a text-book."—*Telegraphic Journal.*

Electricity.

A MANUAL OF ELECTRICITY: Including Galvanism, M netism, Dia-Magnetism, Electro-Dynamics, Magno-Electricity, and the El Telegraph. By HENRY M. NOAD, Ph.D., F.R.S., F.C.S. Fourth Ed With 500 Woodcuts. 8vo, £1 4*s.* cloth.

"The accounts given of electricity and galvanism are not only complete in a scientific but, which is a rarer thing, are popular and interesting."—*Lancet.*

"It is worthy of a place in the library of every public institution."—*Mining Journal.*

Electric Light.

ELECTRIC LIGHT: Its Production and Use. Embodying Pr Directions for the Treatment of Voltaic Batteries, Electric Lamps, Dynamo-Electric Machines. By J. W. URQUHART, C.E., Author of "Elec plating: A Practical Handbook." Edited by F. C. WEBB, M.I.C.E., M.S.T Second Edition, Revised, with large Additions and 128 Illusts. 7*s.* 6*d.* cl

"The book is by far the best that we have yet met with on the subject."—*Athenæum.*

"It is the only work at present available which gives, in language intelligible for the most to the ordinary reader, a general but concise history of the means which have been adopted the present time in producing the electric light."—*Metropolitan*

Electric Lighting.

THE ELEMENTARY PRINCIPLES OF ELECTRIC LIGH ING. By ALAN A. CAMPBELL SWINTON, Associate S.T.E. Crown 1*s.* 6*d.*, cloth. [*Just publis*

"As a stepping-stone to treatises of a more advanced nature, this little work will be most efficient."—*Bookseller.*

"Anyone who desires a short and thoroughly clear exposition of the elementary princip electric-lighting cannot do better than read this little work."—*Bradford Observer.*

Dr. Lardner's School Handbooks.

NATURAL PHILOSOPHY FOR SCHOOLS. By Dr. LARDN 328 Illustrations. Sixth Edition. One Vol., 3*s.* 6*d.* cloth.

"A very convenient class-book for junior students in private schools. It is intended to co in clear and precise terms, general notions of all the principal divisions of Physical Science *British Quarterly Review.*

ANIMAL PHYSIOLOGY FOR SCHOOLS. By Dr. LARDN With 190 Illustrations. Second Edition. One Vol., 3*s.* 6*d.* cloth.

"Clearly written, well a anged, and excellently illustrated."—*Gardener's Chronicle.*

Dr. Lardner's Electric Telegraph.

THE ELECTRIC TELEGRAPH. By Dr. LARDNER. R vised and Re-written by E. B. BRIGHT, F.R.A.S. 140 Illustrations. Sm 8vo, 2*s.* 6*d.* cloth.

"One of the most readable books extant on the Electric Telegraph."—*English Mechanic*

'ms.

ORMS: Their Nature, Classification, and Laws; with the Means Predicting them by their Embodiments, the Clouds. By WILLIAM ASIUS. With Coloured Plates and numerous Wood Engravings. Crown o, 10s. 6d. cloth.

useful repository to meteorologists in the study of atmospherical disturbances. Will repay as being the production of one who gives evidence of acute observation."—*Nature.*

Blowpipe.

E BLOWPIPE IN CHEMISTRY, MINERALOGY, AND OLOGY. Containing all known Methods of Anhydrous Analysis, many orking Examples, and Instructions for Making Apparatus. By Lieut.-lonel W. A. Ross, R.A., F.G.S. With 120 Illustrations. Crown 8vo, 6d. cloth.

ie student who goes conscientiously through the course of experimentation here laid down i a better insight into inorganic chemistry and mineralogy than if he had 'got up' any of text-books of the day, and passed any number of examinations."—*Chemical News.*

Military Sciences.

IDE-MEMOIRE TO THE MILITARY SCIENCES. Framed m Contributions of Officers and others connected with the different Ser-:es. Originally edited by a Committee of the Corps of Royal Engineers. cond Edition, most carefully revised by an Officer of the Corps, with many ditions; containing nearly 350 Engravings and many hundred Woodcuts. iree Vols., royal 8vo, extra cloth boards, and lettered, £4 10s.

compendious encyclopædia of military knowledge, to which we are greatly indebted."—*rgh Review.*

ie most comprehensive work of reference to the military and collateral sciences.'—*Volun-vice Gazette.*

d Fortification.

TREATISE ON FIELD FORTIFICATION, THE ATTACK FORTRESSES, MILITARY MINING, AND RECONNOITRING. By olonel I. S. MACAULAY, late Professor of Fortification in the R.M.A., Wool-ich. Sixth Edition, crown 8vo, cloth, with separate Atlas of 12 Plates, 12s.

,chology.

IANUAL OF THE MOLLUSCA: A Treatise on Recent and ossil Shells. By Dr. S. P. WOODWARD, A.L.S. With Appendix by RALPH 'ATE, A.L.S., F.G.S. With numerous Plates and 300 Woodcuts. Cloth oards, 7s. 6d.

. most valuable storehouse of conchological and geological information."—*Hardwicke's Gossip.*

ronomy.

'STRONOMY. By the late Rev. ROBERT MAIN, M.A., F.R.S., rmerly Radcliffe Observer at Oxford. Third Edition, Revised and Cor-ected to the present time, by WILLIAM THYNNE LYNN, B.A., F.R.A.S., formerly f the Royal Observatory, Greenwich. 12mo, 2s. cloth limp.

. sound and simple treatise, carefully edited, and a capital book for beginners."—*Knowledge.* .ccurately brought down to the requirements of the present time."—*Educational Times.*

logy.

'UDIMENTARY TREATISE ON GEOLOGY, PHYSICAL ND HISTORICAL. Consisting of "Physical Geology," which sets forth ie leading Principles of the Science; and "Historical Geology," which eats of the Mineral and Organic Conditions of the Earth at each successive poch, especial reference being made to the British Series of Rocks. By ALPH TATE, A.L.S., F.G.S., &c., &c. With 250 Illustrations. 12mo, 5s. oth boards.

The fulness of the matter has elevated the book into a manual. Its information is exhaustive ll arranged."—*School Board Chronicle.*

logy and Genesis.

'HE TWIN RECORDS OF CREATION; or, Geology and enesis: their Perfect Harmony and Wonderful Concord. By GEORGE W. 'ICTOR LE VAUX. Numerous Illustrations. Fcap. 8vo, 5s. cloth.

valuable contribution to the evidences of revelation, and disposes very conclusively of the ents of those who would set God's Works against God's Word. No real difficulty is slurked, sophistry is left unexposed."—*The Rock.*

Dr. LARDNER'S HANDBOOKS of NATURAL PHILOSOPHY.

**** The following five volumes, though each is complete in itself, and to be pu-chased separately, form* A COMPLETE COURSE OF NATURAL PHILOSOPHY. *Th- style is studiously popular. It has been the author's aim to supply Manuals for th- Student, the Engineer, the Artisan, and the superior classes in Schools.*

THE HANDBOOK OF MECHANICS. Enlarged and almost re-written by BENJAMIN LOEWY, F.R.A.S. With 378 Illustrations. Post 8v- 6s. cloth.

"The perspicuity of the original has been retained, and chapters which had become obsole- have been replaced by others of more modern character. The explanations throughout a- studiously popular, and care has been taken to show the application of the various branches physics to the industrial arts, and to the practical business of life."—*Mining Journal.*

"Mr. Loewy has carefully revised the book, and brought it up to modern requirements."—*Nature.*

"Natural philosophy has had few exponents more able or better skilled in the art of pop-larising the subject than Dr. Lardner; and Mr. Loewy is doing good service in fitting this treati- and the others of the series, for use at the present time."—*Scotsman.*

THE HANDBOOK OF HYDROSTATICS AND PNEUMATICS. New Edition, Revised and Enlarged, by BENJAMIN LOEWY, F.R.A.S. Wit- 236 Illustrations. Post 8vo, 5s. cloth.

"For those 'who desire to attain an accurate knowledge of physical science without the pr- found methods of mathematical investigation,' this work is not merely intended, but well adapted." —*Chemical News.*

"The volume before us has been carefully edited, augmented to nearly twice the bulk of th- former edition, and all the most recent matter has been added. . . . It is a valuable text-book." —*Nature.*

"Candidates for pass examinations will find it, we think, specially suited to their requirements."—*English Mechanic.*

THE HANDBOOK OF HEAT. Edited and almost entirely re-written by BENJAMIN LOEWY, F.R.A.S., &c. 117 Illustrations. Post 8vo, 6s. cloth.

"The style is always clear and precise, and conveys instruction without leaving any cloudines- or lurking doubts behind."—*Engineering.*

"A most exhaustive book on the subject on which it treats, and is so arranged that it can be understood by all who desire to attain an accurate knowledge of physical science. . . . . Mr. Loewy has included all the latest discoveries in the varied laws and effects of heat."—*Standard.*

"A complete and handy text-book for the use of students and general readers."—*English Mechanic.*

THE HANDBOOK OF OPTICS. By DIONYSIUS LARDNER, D.C.L. formerly Professor of Natural Philosophy and Astronomy in University College, London. New Edition. Edited by T. OLVER HARDING, B.A. Lond., of University College, London. With 298 Illustrations. Small 8vo, 448 pages, 5s. cloth.

"Written by one of the ablest English scientific writers, beautifully and elaborately illustrated." —*Mechanics' Magazine.*

THE HANDBOOK OF ELECTRICITY, MAGNETISM, AND ACOUSTICS. By Dr. LARDNER. Ninth Thousand. Edit. by GEORGE CAREY FOSTER, B.A., F.C.S. With 400 Illustrations. Small 8vo, 5s. cloth.

"The book could not have been entrusted to anyone better calculated to preserve the terse and lucid style of Lardner, while correcting his errors and bringing up his work to the present state of scientific knowledge."—*Popular Science Review.*

Dr. Lardner's Handbook of Astronomy.

THE HANDBOOK OF ASTRONOMY. Forming a Companion to the "Handbook of Natural Philosophy." By DIONYSIUS LARDNER, D.C.L., formerly Professor of Natural Philosophy and Astronomy in University College, London. Fourth Edition. Revised and Edited by EDWIN DUNKIN, F.R.A.S., Royal Observatory, Greenwich. With 38 Plates and upwards of 100 Woodcuts. In One Vol., small 8vo, 550 pages, 9s. 6d. cloth.

"Probably no other book contains the same amount of information in so compendious and well-arranged a form—certainly none at the price at which this is offered to the public."—*Athenæum.*

"We can do no other than pronounce this work a most valuable manual of astronomy, and we strongly recommend it to all who wish to acquire a general—but at the same time correct—acquaintance with this sublime science."—*Quarterly Journal of Science.*

"One of the most deservedly popular books on the subject . . . We would recommend not only the student of the elementary principles of the science, but he who aims at mastering the higher and mathematical branches of astronomy, not to be without this work beside him."—*Practical Magazine.*

DR. LARDNER'S MUSEUM OF SCIENCE AND ART.

THE MUSEUM OF SCIENCE AND ART. Edited DIONYSIUS LARDNER, D.C.L., formerly Professor of Natural Philosophy and A.troı o ·y in University College, London. With upwards of 1,200 Engravings on Wood. In 6 Double Volumes, £1 1s., in a new and elegant cloth binding; or handsomely bound in half-morocco, 31s. 6d.

Contents:

The Planets: Are they Inhabited Worlds?—Weather Prognostics—Popular Fallacies in Questions of Physical Science—Latitudes and Longitudes—Lunar Influences—Meteoric Stones and Shooting Stars—Railway Accidents—Light—Common Things: Air—Locomotion in the United States—Cometary Influences—Common Things: Water—The Potter's Art—Common Things: Fire—Locomotion and Transport, their Influence and Progress—The Moon—Common Things: The Earth—The Electric Telegraph—Terrestrial Heat—The Sun—Earthquakes and Volcanoes—Barometer, Safety Lamp, and Whitworth's Micrometric Apparatus—Steam—The Steam Engine—The Eye—The Atmosphere—Time—Common Things: Pumps—Common Things: Spectacles, the Kaleidoscope—Clocks and Watches—Microscopic Drawing and Engraving—Locomotive—Thermometer—New Planets: Leverrier and Adams's Planet—Magnitude and Minuteness—Common Things: The Almanack—Optical Images—How to observe the Heavens—Common Things: The Looking-glass—Stellar Universe—The Tides—Colour—Common Things: Man—Magnifying Glasses—Instinct and Intelligence—The Solar Microscope—The Camera Lucida—The Magic Lantern—The Camera Obscura—The Microscope—The White Ants: Their Manners and Habits—The Surface of the Earth, or First Notions of Geography—Science and Poetry—The Bee—Steam Navigation—Electro-Motive Power—Thunder, Lightning, and the Aurora Borealis—The Printing Press—The Crust of the Earth—Comets—The Stereoscope—The Pre-Adamite Earth—Eclipses—Sound.

Opinions of the Press.

"This series, besides affording popular but sound instruction on scientific subjects, with which the humblest man in the country ought to be acquainted, also undertakes that teaching of 'Common Things' which every well-wisher of his kind is anxious to promote. Many thousand copies of this serviceable publication have been printed, in the belief and hope that the desire for instruction and improvement widely prevails; and we have no fear that such enlightened faith will meet with disappointment."—*Times.*

"A cheap and interesting publication, alike informing and attractive. The papers combine subjects of importance and great scientific knowledge, considerable inductive powers, and a popular style of treatment."—*Spectator.*

"The 'Museum of Science and Art' is the most valuable contribution that has ever been made to the Scientific Instruction of every class of society."—Sir DAVID BREWSTER, in the *North British Review.*

"Whether we consider the liberality and beauty of the illustrations, the charm of the writing, or the durable interest of the matter, we must express our belief that there is hardly to be found among the new books one that would be welcomed by people of so many ages and classes as a valuable present."—*Examiner.*

*** *Separate books formed from the above, suitable for Workmen's Libraries, Science Classes, &c.*

Common Things Explained. Containing Air, Earth, Fire, Water, Time, Man, the Eye, Locomotion, Colour, Clocks and Watches, &c. 233 Illustrations, cloth gilt, 5s.

The Microscope. Containing Optical Images, Magnifying Glasses, Origin and Description of the Microscope, Microscopic Objects, the Solar Microscope, Microscopic Drawing and Engraving, &c. 147 Illustrations, cloth gilt, 2s.

Popular Geology. Containing Earthquakes and Volcanoes, the Crust the Earth, &c. 201 Illustrations, cloth gilt, 2s. 6d.

Popular Physics. Containing Magnitude and Minuteness the ı sphere, Meteoric Stones, Popular Fallacies, Weather Prognostics, Thermometer, the Barometer, Sound, &c. 85 Illustrations, cloth gilt, 2s.

Steam and its Uses. Including the Steam Engine, the Locomotive, an Steam Navigation. 89 Illustrations, cloth gilt, 2s.

Popular Astronomy. Containing How to observe the Heavens—Th Earth, Sun, Moon, Planets, Light, Comets, Eclipses, Astronomical Influences, &c. 182 Illustrations, 4s. 6d.

The Bee and White Ants: Their Manners and Habits. With Illustrations of Animal Instinct and Intelligence. 135 Illustrations, cloth gilt, 2s.

The Electric Telegraph Popularised. To render intelligible to all who can Read, irrespective of any previous Scientific Acquirements, the various forms of Telegraphy in Actual Operation. 100 Illustrations, cloth gilt, 1s. 6d.

MATHEMATICS, GEOMETRY, TABLES, etc.

Practical Mathematics.

MATHEMATICS FOR PRACTICAL MEN. Being a Common-place Book of Pure and Mixed Mathematics. Designed chiefly for the Use of Civil Engineers, Architects and Surveyors. By OLINTHUS GREGORY, LL.D., F.R.A.S., Enlarged by HENRY LAW, C.E. 4th Edition, carefully Revised by J. R. YOUNG, formerly Professor of Mathematics, Belfast College. With 13 Plates, 8vo, £1 1s. cloth.

"The engineer or architect will here find ready to his hand rules for solving nearly every mathematical difficulty that may arise in his practice. The rules are in all cases explained by means of examples, in which every step of the process is clearly worked out."—*Builder.*

"One of the most serviceable books for practical mechanics. . . . It is an instructive book for the student, and a Text-book for him who, having once mastered the subjects it treats of, needs occasionally to refresh his memory upon them."—*Building News.*

Metrical Units and Systems, etc.

MODERN METROLOGY: A Manual of the Metrical Units and Systems of the Present Century. With an Appendix containing a proposed English System. By LOWIS D'A. JACKSON, A.M. Inst. C.E., Author of "Aid to Survey Practice," &c. Large crown 8vo, 12s. 6d. cloth.

"The author has brought together much valuable and interesting information. . . . We cannot but recommend the work to the consideration of all interested in the practical reform of our weights and measures."—*Nature.*

"For exhaustive tables of equivalent weights and measures of all sorts, and for clear demonstrations of the effects of the various systems that have been proposed or adopted, Mr. Jackson's treatise is without a rival."—*Academy.*

The Metric System.

A SERIES OF METRIC TABLES, in which the British Standard Measures and Weights are compared with those of the Metric System at present in Use on the Continent. By C. H. DOWLING, C.E. 8vo, 10s. 6d. strongly bound.

"Their accuracy has been certified by Professor Airy, the Astronomer-Royal."—*Builder.*

"Mr. Dowling's Tables are well put together as a ready-reckoner for the conversion of one system into the other."—*Athenæum*

Geometry for the Architect, Engineer, etc.

PRACTICAL GEOMETRY, for the Architect, Engineer and Mechanic. Giving Rules for the Delineation and Application of various Geometrical Lines, Figures and Curves. By E. W. TARN, M.A., Architect, Author of "The Science of Building," &c. Second Edition. With Appendices on Diagrams of Strains and Isometrical Projection. With 172 Illustrations, demy 8vo, 9s. cloth.

"No book with the same objects in view has ever been published in which the clearness of the rules laid down and the illustrative diagrams have been so satisfactory."—*Scotsman.*

"This is a manual for the practical man, whether architect, engineer, or mechanic. . . . The object of the author being to avoid all abstruse formulæ or complicated methods, and to enable persons with but a moderate knowledge of geometry to work out the problems required."—*English Mechanic.*

The Science of Geometry.

THE GEOMETRY OF COMPASSES; or, Problems Resolved by the mere Description of Circles and the use of Coloured Diagrams and Symbols. By OLIVER BYRNE. Coloured Plates. Crown 8vo, 3s. 6d. cloth.

"The treatise is a good one, and remarkable—like all Mr. Byrne's contributions to the science of geometry—for the lucid character of its teaching."—*Building News.*

Iron and Metal Trades' Calculator.

THE IRON AND METAL TRADES' COMPANION. For expeditiously ascertaining the Value of any Goods bought or sold by Weight, from 1s. per cwt. to 112s. per cwt., and from one farthing per pound to one shilling per pound. Each Table extends from one pound to 100 tons. To which are appended Rules on Decimals, Square and Cube Root, Mensuration of Superficies and Solids, &c.; Tables of Weights of Materials, and other Useful Memoranda. By THOS. DOWNIE. 396 pp., 9s. Strongly bound leather.

"A most useful set of tables, and will supply a want, for nothing like them before existed."—*Building News.*

"Although specially adapted to the iron and metal trades, the tables will be found useful in every other business in which merchandise is bought and sold by weight."—*Railway News.*

Calculator for Numbers and Weights Combined.

THE COMBINED NUMBER AND WEIGHT CALCULATOR. Containing upwards of 250,000 Separate Calculations, showing at a glance the value at 421 different rates, ranging from $\frac{1}{128}$th of a Penny to 20s. each, or per cwt., and £20 per ton, of any number of articles consecutively, from 1 to 470.—Any number of cwts., qrs., and lbs., from 1 cwt. to 470 cwts.—Any number of tons, cwts., qrs., and lbs., from 1 to 23½ tons. By WILLIAM CHADWICK, Public Accountant. Imp. 8vo, 30s., strongly bound.

☞ *This comprehensive and entirely unique and original Calculator is adapted for the use of Accountants and Auditors, Railway Companies, Canal Companies, Shippers, Shipping Agents, General Carriers, &c.*

Ironfounders, Brassfounders, Metal Merchants, Iron Manufacturers, Ironmongers, Engineers, Machinists, Boiler Makers, Millwrights, Roofing, Bridge and Girder Makers, Colliery Proprietors, &c.

Timber Merchants, Builders, Contractors, Architects, Surveyors, Auctioneers, Valuers, Brokers, Mill Owners and Manufacturers, Mill Furnishers, Merchants and General Wholesale Tradesmen.

*** OPINIONS OF THE PRESS.

"The book contains the answers to questions, and not simply a set of ingenious puzzle methods of arriving at results. It is as easy of reference for any answer or any number of answers as a dictionary, and the references are even more quickly made. For making up accounts or estimates, the book must prove invaluable to all who have any considerable quantity of calculations involving price and measure in any combination to do."—*Engineer.*

"The most complete and practical ready reckoner which it has been our fortune yet to see. It is difficult to imagine a trade or occupation in which it could not be of the greatest use, either in saving human labour or in checking work. The publishers have placed within the reach of every commercial man an invaluable and unfailing assistant."—*The Miller.*

Comprehensive Weight Calculator.

THE WEIGHT CALCULATOR. Being a Series of Tables upon a New and Comprehensive Plan, exhibiting at One Reference the exact Value of any Weight from 1 lb. to 15 tons, at 300 Progressive Rates, from 1*d.* to 168*s.* per cwt., and containing 186,000 Direct Answers, which, with their Combinations, consisting of a single addition (mostly to be performed at sight), will afford an aggregate of 10,266,000 Answers; the whole being calculated and designed to ensure correctness and promote despatch. By HENRY HARBEN, Accountant. An entirely New Edition, carefully Revised. Royal 8vo, strongly half-bound, £1 5*s.*

"Of priceless value to business men. Its accuracy and completeness have secured for it a reputation which renders it quite unnecessary for us to say one word in its praise. It is a necessary book in all mercantile offices."—*Sheffield Independent.*

Comprehensive Discount Guide.

THE DISCOUNT GUIDE. Comprising several Series of Tables for the use of Merchants, Manufacturers, Ironmongers, and others, by which may be ascertained the exact Profit arising from any mode of using Discounts, either in the Purchase or Sale of Goods, and the method of either Altering a Rate of Discount or Advancing a Price, so as to produce, by one operation, a sum that will realise any required profit after allowing one or more Discounts: to which are added Tables of Profit or Advance from 1¼ to 90 per cent., Tables of Discount from 1¼ to 98¾ per cent., and Tables of Commission, &c., from ⅛ to 10 per cent. By HENRY HARBEN, Accountant, Author of "The Weight Calculator." New Edition, carefully Revised and Corrected. Demy 8vo, 544 pp. half-bound, £1 5*s.*

"A book such as this can only be appreciated by business men, to whom the saving of time means saving of money. We have the high authority of Professor J. R. Young that the tables throughout the work are constructed upon strictly accurate principles. The work must prove of great value to merchants, manufacturers, and general traders."—*British Trade Journal*

Iron Shipbuilders' and Iron Merchants' Tables.

IRON-PLATE WEIGHT TABLES: For Iron Shipbuilders, Engineers and Iron Merchants. Containing the Calculated Weights of upwards of 150,000 different sizes of Iron Plates, from 1 foot by 6 in. by ¼ in. to 10 feet by 5 feet by 1 in. Worked out on the basis of 40 lbs. to the square foot of Iron of 1 inch in thickness. Carefully compiled and thoroughly Revised by H. BURLINSON and W. H. SIMPSON. Oblong 4to, 25*s.* half-bound.

"This work will be found of great utility. The authors have had much practical experience of what is wanting in making estimates; and the use of the book will save much time in making elaborate calculations."—*English Mechanic.*

INDUSTRIAL AND USEFUL ARTS.

Soap-making.

THE ART OF SOAP-MAKING: A Practical Handbook of the Manufacture of Hard and Soft Soaps, Toilet Soaps, &c. Including many New Processes, and a Chapter on the Recovery of Glycerine from Waste Leys. By ALEXANDER WATT, Author of "Electro-Metallurgy Practically Treated," &c. With numerous Illustrations. Second Edition, Revised. Crown 8vo, 9s. cloth.

"The work will prove very useful, not merely to the technological student, but to the practical soapboiler who wishes to understand the theory of his art."—*Chemical News.*

"It is really an excellent example of a technical manual, entering, as it does, thoroughly and exhaustively both into the theory and practice of soap manufacture."—*Knowledge.*

"Mr. Watt's book is a thoroughly practical treatise on an art which has almost no literature in our language We congratulate the author on the success of his endeavour to fill a void in English technical literature."—*Nature.*

Leather Manufacture.

THE ART OF LEATHER MANUFACTURE. Being a Practical Handbook, in which the Operations of Tanning, Currying, and Leather Dressing are fully Described, and the Principles of Tanning Explained, and many Recent Processes introduced; as also Methods for the Estimation of Tannin, and a Description of the Arts of Glue Boiling, Gut Dressing, &c. By ALEXANDER WATT, Author of "Soap-Making," "Electro-Metallurgy," &c. With numerous Illustrations. Crown 8vo, 12s. 6d. cloth.

"Mr. Watt has rendered an important service to the trade, and no less to the student of technology."—*Chemical News.*

"A sound, comprehensive treatise. The book is an eminently valuable production which redounds to the credit of both author and publishers."—*Chemical Review.*

"This volume is technical without being tedious, comprehensive and complete without being prosy, and it bears on every page the impress of a master hand. We have never come across a better trade treatise, nor one that so thoroughly supplied an absolute want."—*Shoe and Leather Trades' Chronicle.*

Boot and Shoe Making.

THE ART OF BOOT AND SHOE-MAKING. A Practical Handbook, including Measurement, Last-Fitting, Cutting-Out, Closing and Making, with a Description of the most approved Machinery employed. By JOHN B. LENO, late Editor of *St. Crispin*, and *The Boot and Shoe-Maker*. With numerous Illustrations. Crown 8vo, 5s. cloth.

"This excellent treatise is by far the best work ever written on the subject. A new work, embracing all modern improvements, was much wanted. This want is now satisfied. The chapter on clicking, which shows how waste may be prevented, will save fifty times the price of the book."—*Scottish Leather Trader.*

Dentistry.

MECHANICAL DENTISTRY: A Practical Treatise on the Construction of the various kinds of Artificial Dentures. Comprising also Useful Formulæ, Tables and Receipts for Gold Plate, Clasps, Solders, &c. &c. By CHARLES HUNTER. Second Edition, Revised. With upwards of 100 Wood Engravings. Crown 8vo, 7s. 6d. cloth.

"We can strongly recommend Mr. Hunter's treatise to all students preparing for the profession of dentistry, as well as to every mechanical dentist."—*Dublin Journal of Medical Science.*

"A work in a concise form that few could read without gaining information from."—*British Journal of Dental Science.*

Brewing.

A HANDBOOK FOR YOUNG BREWERS. By HERBERT EDWARDS WRIGHT, B.A. Crown 8vo, 3s. 6d. cloth.

"This little volume, containing such a large amount of good sense in so small a compass, ought to recommend itself to every brewery pupil, and many who have passed that stage."—*Brewers' Guardian.*

"The book is very clearly written, and the author has successfully brought his scientific knowledge to bear upon the various processes and details of brewing."—*Brewer.*

Wood Engraving.

A PRACTICAL MANUAL OF WOOD ENGRAVING. With a Brief Account of the History of the Art. By WILLIAM NORMAN BROWN. With numerous Illustrations. Crown 8vo, 2s. cloth.

"The author deals with the subject in a thoroughly practical and easy series of representative lessons."—*Paper and Printing Trades' Journal.*

Electrolysis of Gold, Silver, Copper, &c.

ELECTRO-DEPOSITION: A Practical Treatise on the Electrolysis of Gold, Silver, Copper, Nickel, and other Metals and Alloys. With descriptions of Voltaic Batteries, Magnets and Dynamo-Electric Machines, Thermopiles, and of the Materials and Processes used in every Department of the Art, and several Chapters on ELECTRO-METALLURGY. By ALEXANDER WATT, Author of "Electro-Metallurgy," "The Art of Soapmaking," &c. With numerous Illustrations. Crown 8vo, 12s. 6d., cloth.

"Evidently written by a practical man who has spent a long period of time in electro-plate workshops. The information given respecting the details of workshop manipulation is remarkably complete. . . . Mr. Watt's book will prove of great value to electro-depositors, jewellers, and various other workers in metal."—*Nature.*

"Eminently a book for the practical worker in electro-deposition. It contains minute and practical descriptions of methods, processes and materials as actually pursued and used in the workshop. Mr. Watt's book recommends itself to all interested in its subjects.—*Engineer.*

"Contains an enormous quantity of practical information; and there are probably few items omitted which could be of any possible utility to workers in galvano-plasty. As a practical manual the book can be recommended to all who wish to study the art of electro-deposition."—*English Mechanic.*

Electroplating, etc.

ELECTROPLATING: A Practical Handbook. By J. W. URQUHART, C.E. With numerous Illustrations. Crown 8vo, 5s. cloth.

"The information given appears to be based on direct personal knowledge. . . Its science sound and the style is always clear."—*Athenæum.*

Electrotyping, etc.

ELECTROTYPING: The Reproduction and Multiplication of Printing Surfaces and Works of Art by the Electro-deposition of Metals. By J. W. URQUHART, C.E. Crown 8vo, 5s. cloth.

"The book is thoroughly practical. The reader is, therefore, conducted through the leading laws of electricity, then through the metals used by electrotypers, the apparatus, and the depositing processes, up to the final preparation of the work."—*Art Journal.*

"We can recommend this treatise, not merely to amateurs, but to those actually engaged in the trade."—*Chemical News.*

Electro-Metallurgy.

ELECTRO-METALLURGY; Practically Treated. By ALEXANDER WATT, F.R.S.S.A. Eighth Edition, Revised, with Additional Matter and Illustrations, including the most recent Processes. 12mo, 3s. 6d. cloth boards.

"From this book both amateur and artisan may learn everything necessary for the successful prosecution of electroplating."—*Iron.*

Goldsmiths' Work.

THE GOLDSMITH'S HANDBOOK. By GEORGE E. GEE, Jeweller, &c. Third Edition, considerably Enlarged. 12mo, 3s. 6d. cloth boards.

"A good, sound, technical educator, and will be generally accepted as an authority. It is essentially a book for the workshop, and exactly fulfils the purpose intended."—*Horological Journal.*

"Will speedily become a standard book which few will care to be without."—*Jeweller and Metalworker.*

Silversmiths' Work.

THE SILVERSMITH'S HANDBOOK. By GEORGE E. GEE, Jeweller, &c. Second Edition, Revised, with numerous Illustrations. 12mo 3s. 6d. cloth boards.

"The chief merit of the work is its practical character. . . The workers in the trade will speedily discover its merits when they sit down to study it."—*English Mechanic.*

"This work forms a valuable sequel to the author's 'Goldsmith's Handbook.'"—*Silversmiths' Trade Journal.*

*** *The above two works together, strongly half-bound, price 7s.*

Textile Manufacturers' Tables.

UNIVERSAL TABLES OF TEXTILE STRUCTURE. For the use of Manufacturers in every branch of Textile Trade. By JOSEPH EDMONDSON. Oblong folio, strongly bound in cloth, price 7s. 6d.

☞ *These Tables provide what has long been wanted, a simple and easy means of adjusting yarns to "reeds," or "setts," or to "picks" or "shots," and vice versa, so that fabrics may be made of varying weights or fineness, but having the same character and proportions.*

;HEMICAL MANUFACTURES & COMMERCE.

.The Alkali Trade, Manufacture of Sulphuric Acid, etc.

A MANUAL OF THE ALKALI TRADE, including the Manufacture of Sulphuric Acid, Sulphate of Soda, and Bleaching Powder. By JOHN LOMAS, Alkali Manufacturer, Newcastle-upon-Tyne and London. With 232 Illustrations and Working Drawings, and containing 390 pages of Text. Second Edition, with Additions. Super-royal 8vo, £1 10s. cloth.

*** This work provides (1) a Complete Handbook for intending Alkali and ılphuric Acid Manufacturers, and for those already in the field who desire to ıprove their plant, or to become practically acquainted with the latest processes ·d developments of the trade: (2) a Handy Volume which Manufacturers can t into the hands of their Managers and Foremen as a useful guide in their daily unds of duty.

"The author has given the fullest, most practical, and, to all concerned in the alkali trade, most uable mass of information that, to our knowledge, has been published."—*Engineer.*

"This book is written by a manufacturer for manufacturers. The working details of the most ›roved forms of apparatus are given, and these are accompanied by no less than 232 wood en-vings, all of which may be used for the purposes of construction. Every step in the manufac-e is very fully described in this manual, and each improvement explained. Everything which ds to introduce economy into the technical details of this trade receives the fullest attention."—*'enæum.*

The author is not one of those clever compilers who, on short notice, will 'read up' any conceiv-e subject, but a practical man in the best sense of the word. We find here not merely a sound l luminous explanation of the chemical principles of the trade, but a notice of numerous matters ch have a most important bearing on the successful conduct of alkali works, but which are .erally overlooked by even the most experienced technological authors."—*Chemical Review.*

›mmercial Chemical Analysis.

THE COMMERCIAL HANDBOOK OF CHEMICAL ANALYSIS; or, Practical Instructions for the determination of the Intrinsic or Commercial Value of Substances used in Manufactures, in Trades, and in the Arts. By A. NORMANDY, Editor of Rose's "Treatise on Chemical Analysis." New Edition, to a great extent Re-written, by HENRY M. NOAD, Ph.D., F.R.S. With numerous Illustrations. Crown 8vo, 12s. 6d. cloth.

"We strongly recommend this book to our readers as a guide, alike indispensable to the house-: as to the pharmaceutical practitioner."—*Medical Times.*

"Essential to the analysts appointed under the new Act. The most recent results are given, the work is well edited and carefully written."—*Nature.*

ye-Wares and Colours.

THE MANUAL OF COLOURS AND DYE-WARES: Their Properties, Applications, Valuation, Impurities, and Sophistications. For the use of Dyers, Printers, Drysalters, Brokers, &c. By J. W. SLATER. Second Edition, Revised and greatly Enlarged. Crown 8vo, 7s. 6d. cloth.

"A complete encyclopædia of the *materia tinctoria*. The information given respecting each le is full and precise, and the methods of determining the value of articles such as these, so e to sophistication, are given with clearness, and are practical as well as valuable."—*Chemist Druggist.*

"There is no other work which covers precisely the same ground. To students preparing :xaminations in dyeing and printing it will prove exceedingly useful."—*Chemical News.*

'gments.

THE ARTIST'S MANUAL OF PIGMENTS. Showing their Composition, Conditions of Permanency, Non-Permanency, and Adulterations; Effects in Combination with Each Other and with Vehicles; and the most Reliable Tests of Purity. Together with the Science and Arts Department's Examination Questions on Painting. By H. C. STANDAGE. Small crown 8vo, 2s. 6d. cloth.

"This work is indeed *multum-in-parvo*, and we can, with good conscience, recommend it to ho come in contact with pigments, whether as makers, dealers or users."—*Chemical Review.*

'This manual cannot fail to be a very valuable aid to all painters who wish their work to ire and be of a sound character; it is complete and comprehensive."—*Spectator.*

'The author supplies a great deal of very valuable information and memoranda as to the .ıical qualities and artistic effect of the Principal pigments used by painters."—*Builder.*

AGRICULTURE, LAND MANAGEMENT, etc.

Youatt and Burn's Complete Grazier.

THE COMPLETE GRAZIER, and FARMER'S and CATTLE BREEDER'S ASSISTANT. A Compendium of Husbandry; especially i the departments connected with the Breeding, Rearing, Feeding, and Gener: Management of Stock; the Management of the Dairy, &c. With Directior for the Culture and Management of Grass Land, of Grain and Root Crop the Arrangement of Farm Offices, the use of Implements and Machines, an on Draining, Irrigation, Warping, &c.; and the Application and Relativ Value of Manures. By WILLIAM YOUATT, Esq., V.S. Twelfth Edition, E larged, by ROBERT SCOTT BURN, Author of "Outlines of Modern Farming "Systematic Small Farming," &c. One large 8vo Volume, 860 pp., with 24 Illustrations, £1 1s. half-bound.

"The standard and text-book with the farmer and grazier."—*Farmers' Magazine.*

"A treatise which will remain a standard work on the subject as long as British agricultu endures."—*Mark Lane Express* (First Notice).

The book deals with all departments of agriculture, and contains an immense amount valuable information. It is, in fact, an encyclopædia of agriculture put into readable form, and is the only work equally comprehensive brought down to present date. It deserves a place in tl library of every agriculturist."—*Mark Lane Express* (Second Notice)

"This esteemed work is well worthy of a place in the libraries of agriculturists."—*Nor British Agriculturist.*

Modern Farming.

OUTLINES OF MODERN FARMING. By R. SCOTT BURN Soils, Manures, and Crops—Farming and Farming Economy—Cattle, Shee and Horses—Management of the Dairy, Pigs and Poultry—Utilisation Town-Sewage, Irrigation, &c. Sixth Edition. In One Vol., 1,250 pp., ha bound, profusely Illustrated, 12s.

"The aim of the author has been to make his work at once comprehensive and trustworth and in this aim he has succeeded to a degree which entitles him to much credit."—*Morn Advertiser.*

"Eminently calculated to enlighten the agricultural community on the varied subjects which it treats, and hence it should find a place in every farmer's library."—*City Press.*

Small Farming.

SYSTEMATIC SMALL FARMING; or, The Lessons of Farm. Being an Introduction to Modern Farm Practice for Small Farm in the Culture of Crops; The Feeding of Cattle; The Management of t Dairy, Poultry and Pigs; The Keeping of Farm Work Records; The Ensila System, Construction of Silos, and other Farm Buildings; The Impro ment of Neglected Farms, &c. By ROBERT SCOTT BURN, Author of "O lines of Landed Estates' Management," and "Outlines of Farm Mana ment," and Editor of "The Complete Grazier." With numerous Illustratio crown 8vo, 6s. cloth. [*Just publish*

"This is the completest book of its class we have seen, and one which every amateur farr will read with pleasure and accept as a guide."—*Field.*

"Mr. Scott Burn's pages are severely practical, and the tone of the practical man is throughout. The book can only prove a treasure of aid and suggestion to the small farme intelligence and energy."—*British Quarterly Review,*

Agricultural Engineering.

FARM ENGINEERING, THE COMPLETE TEXT-BOOK O Comprising Draining and Embanking; Irrigation and Water Supply; Fa Roads, Fences, and Gates; Farm Buildings, their arrangement and c struction, with plans and estimates; Barn Implements and Machines; Fi Implements and Machines; Agricultural Surveying, Levelling, &c. By P JOHN SCOTT, Editor of the *Farmers' Gazette,* late Professor of Agricult and Rural Economy at the Royal Agricultural College, Cirencester, &c. In One Vol., 1,150 pages, half-bound, with over 600 Illustrations, 12s.

"Written with great care, as well as with knowledge and ability. The author has done work well; we have found him a very trustworthy guide wherever we have tested his stateme The volume will be of great value to agricultural students, and we have much pleasure in rec mending it."—*Mark Lane Express.*

"For a young agriculturist we know of no handy volume so likely to be more usefully studi —*Bell's Weekly Messenger.*

English Agriculture.

THE FIELDS OF GREAT BRITAIN: A Text-Book of Agriculture, adapted to the Syllabus of the Science and Art Department. For Elementary and Advanced Students. By HUGH CLEMENTS (Board of Trade). 18mo, 2s. 6d. cloth.

"A most comprehensive volume, giving a mass of information."—*Agricultural Economist.*

"It is a long time since we have seen a book which has pleased us more, or which contains such a vast and useful fund of knowledge."—*Educational Times.*

Hudson's Land Valuer's Pocket-Book.

THE LAND VALUER'S BEST ASSISTANT: Being Tables on a very much Improved Plan, for Calculating the Value of Estates. With Tables for reducing Scotch, Irish, and Provincial Customary Acres to Statute Measure, &c. By R. HUDSON, C.E. New Edition. Royal 32mo, leather, elastic band, 4s.

"This new edition includes tables for ascertaining the value of leases for any term of years; and for showing how to lay out plots of ground of certain acres in forms, square, round, &c., with valuable rules for ascertaining the probable worth of standing timber to any amount; and is of incalculable value to the country gentleman and professional man."—*Farmers' Journal.*

Ewart's Land Improver's Pocket-Book.

THE LAND IMPROVER'S POCKET-BOOK OF FORMULÆ, TABLES and MEMORANDA required in any Computation relating to the Permanent Improvement of Landed Property. By JOHN EWART, Land Surveyor and Agricultural Engineer. Second Edition, Revised. Royal 32mo, oblong, leather, gilt edges, with elastic band, 4s.

"A compendious and handy little volume."—*Spectator.*

Complete Agricultural Surveyor's Pocket-Book.

THE LAND VALUER'S AND LAND IMPROVER'S COMPLETE POCKET-BOOK. Consisting of the above Two Works bound together. Leather, gilt edges, with strap, 7s. 6d.

"Hudson's book is the best ready-reckoner on matters relating to the valuation of land and crops, and its combination with Mr. Ewart's work greatly enhances the value and usefulness of the latter-mentioned. . . . It is most useful as a manual for reference."—*North of England Farmer.*

Farm and Estate Book-keeping.

BOOK-KEEPING FOR FARMERS & ESTATE OWNERS. A Practical Treatise, presenting, in Three Plans, a System adapted to all Classes of Farms. By JOHNSON M. WOODMAN, Chartered Accountant. Crown 8vo, 3s. 6d. cloth.

"Will be found of great assistance by those who intend to commence a system of book-keeping, the author's examples being clear and explicit, and his explanations, while full and accurate, being to a large extent free from technicalities."—*Live Stock Journal.*

"The young farmer, land agent and surveyor will find Mr. Woodman's treatise more than repay its cost and study."—*Building News.*

WOODMAN'S YEARLY FARM ACCOUNT BOOK. Giving a Weekly Labour Account and Diary, and showing the Income and Expenditure under each Department of Crops, Live Stock, Dairy, &c. &c. With Valuation, Profit and Loss Account, and Balance Sheet at the end of the Year, and an Appendix of Forms. Folio, 7s. 6d. half-bound.

"Contains every requisite form for keeping farm accounts readily and accurately."—*Agriculture.*

GARDENING, FLORICULTURE, etc.

Early Fruits, Flowers and Vegetables.

THE FORCING GARDEN; or, How to Grow Early Fruits, Flowers, and Vegetables. With Plans, and Estimates for Building Glass-houses, Pits and Frames. Containing also Original Plans for Double Glazing a New Method of Growing the Gooseberry under Glass, &c. &c., and on Ventilation, Protecting Vine Borders, &c. With Illustrations. By SAMUEL WOOD. Crown 8vo, 3s. 6d. cloth.

"A good book, and fairly fills a place that was in some degree vacant."—*Gardeners' Magazine.*

"Mr. Wood's book is an original and exhaustive answer to the question 'How to Grow Early Fruits, Flowers and Vegetables?'"—*Land and Water.*

Good Gardening.

A PLAIN GUIDE TO GOOD GARDENING; or, How to Grow Vegetables, Fruits, and Flowers. With Practical Notes on Soils, Manures, Seeds, Planting, Laying-out of Gardens and Grounds, &c. By S. WOOD. Third Edition, with considerable Additions, &c., and numerous Illustrations. Crown 8vo, 5*s*. cloth.

"A very good book, and one to be highly recommended as a practical guide."—*Athenæum.*

"May be recommended to young gardeners, cottagers, and specially to amateurs, for the plain and trustworthy information it gives on matters too often neglected."—*Gardeners' Chronicle.*

Gainful Gardening.

MULTUM-IN-PARVO GARDENING; or, How to make One Acre of Land produce £620 a-year by the Cultivation of Fruits and Vegetables; also, How to Grow Flowers in Three Glass Houses, so as to realise £176 per annum clear Profit. By SAMUEL WOOD, Author of "Good Gardening," &c. Fourth and cheaper Edition, Revised, with Additions. Crown 8vo, 1*s*. sewed.

"We are bound to recommend it as not only suited to the case of the amateur and gentleman's gardener, but to the market grower."—*Gardeners' Magazine.*

Gardening for Ladies.

THE LADIES' MULTUM-IN-PARVO FLOWER GARDEN, and Amateurs' Complete Guide. With Illustrations. By SAMUEL WOOD. Crown 8vo, 3*s*. 6*d*. cloth.

"This volume contains a good deal of sound, common-sense instruction."—*Florist.*

"Full of shrewd hints and useful instructions, based on a lifetime of experience."—*Scotsman.*

Receipts for Gardeners.

GARDEN RECEIPTS. Edited by CHARLES W. QUIN. 12mo, 1*s*. 6*d*. cloth limp.

"A useful and handy book, containing a good deal of valuable information."—*Athenæum.*

Kitchen Gardening.

THE KITCHEN AND MARKET GARDEN. By Contributors to "The Garden." Compiled by C. W. SHAW, Editor of "Gardening Illustrated." 12mo, 3*s*. 6*d*. cloth boards.

"The most valuable compendium of kitchen and market-garden work published."—*Farmer.*

Cottage Gardening.

COTTAGE GARDENING; or, Flowers, Fruits, and Vegetables for Small Gardens. By E. HOBDAY. 12mo, 1*s*. 6*d*. cloth limp.

"Contains much useful information at a small charge."—*Glasgow Herald.*

AUCTIONEERING, ESTATE AGENCY, etc.

Auctioneer's Assistant.

THE APPRAISER, AUCTIONEER, BROKER, HOUSE AND ESTATE AGENT AND VALUER'S POCKET ASSISTANT, for the Valuation for Purchase, Sale, or Renewal of Leases, Annuities and Reversions, and of property generally; with Prices for Inventories, &c. By JOHN WHEELER. Fifth Edition, Extended by C. NORRIS, Valuer, &c. Royal 32mo, 5*s*. cloth.

"Contains a large quantity of varied and useful information as to the valuation for purchase, sale, or renewal of leases, annuities and reversions, and of property generally, with prices for inventories, and a guide to determine the value of interior fittings and other effects."—*Builder.*

Auctioneering.

AUCTIONEERS: Their Duties and Liabilities. By ROBERT SQUIBBS, Auctioneer. Demy 8vo, 10*s*. 6*d*. cloth.

"The position and duties of auctioneers treated compendiously and clearly."—*Builder.*

"Every auctioneer ought to possess a copy of this excellent work."—*Ironmonger*

How to Invest.

HINTS FOR INVESTORS: Being an Explanation of the Mode of Transacting Business on the Stock Exchange. To which are added Comments on the Fluctuations and Table of Quarterly Average prices of Consols since 1759. Also a Copy of the London Daily Stock and Share List. By WALTER M. PLAYFORD, Sworn Broker. Crown 8vo, 2*s*. cloth.

"An invaluable guide to investors and speculators."—*Bullionist.*

A Complete Epitome of the Laws of this Country.

EVERY MAN'S OWN LAWYER: A Handy-book of the Principles of Law and Equity. By A BARRISTER. Twenty-third Edition. Carefully Revised and brought down to the end of the last Session, including Summaries of the Latest Statute Laws. With Notes and References to the Authorities. Crown 8vo, price 6s. 8*d*. (saved at every consultation), strongly bound in cloth.

Comprising THE RIGHTS AND WRONGS OF INDIVIDUALS—MERCANTILE AND COMMERCIAL LAW—CRIMINAL LAW—PARISH LAW—COUNTY COURT LAW—GAME AND FISHERY LAWS—POOR MEN'S LAWSUITS—THE LAWS OF BANKRUPTCY—BETS AND WAGERS—CHEQUES, BILLS, AND NOTES—CONTRACTS AND AGREEMENTS—COPYRIGHT—ELECTIONS AND REGISTRATION—INSURANCE—LIBEL AND SLANDER—MARRIAGE AND DIVORCE—MERCHANT SHIPPING—MORTGAGES—SETTLEMENTS—STOCK EXCHANGE PRACTICE—TRADE MARKS AND PATENTS—TRESPASS—NUISANCES, &c.—TRANSFER OF LAND, &c.—WARRANTY—WILLS AND AGREEMENTS, &c. &c.

Opinions of the Press.

"*No Englishman ought to be without this book.* . . . Any person perfectly uninformed on legal matters, who may require sound information on unknown law points, will, by reference to this book, acquire the necessary information, and thus on many occasions save the expense and loss of time of a visit to a lawyer."—*Engineer.*

"It is a complete code of English Law, written in plain language, which all can understand."—*Weekly Times.*

"A useful and concise epitome of the law, compiled with considerable care."—*Law Magazine.*

"What it professes to be—a complete epitome of the laws of this country, thoroughly intelligible to non-professional readers. The book is a handy one to have in readiness when some knotty point requires ready solution."—*Bell's Life.*

Metropolitan Rating Appeals.

REPORTS OF APPEALS HEARD BEFORE THE COURT OF GENERAL ASSESSMENT SESSIONS, from the Year 1871 to 1885. By EDWARD RYDE and ARTHUR LYON RYDE. Fourth Edition, brought down to the Present Date, with an Introduction to the Valuation (Metropolis) Act, 1869, and an Appendix by WALTER C. RYDE, of the Inner Temple, Barrister-at-Law. 8vo, 16s. cloth.

House Property.

HANDBOOK OF HOUSE PROPERTY: A Popular and Practical Guide to the Purchase, Mortgage, Tenancy, and Compulsory Sale of Houses and Land. By E. L. TARBUCK, Architect and Surveyor. Third Edition, 12mo, 3s. 6*d*. cloth.

"The advice is thoroughly practical."—*Law Journal.*

"This is a well-written and thoughtful work. We commend the work to the careful study of all interested in questions affecting houses and land."—*Land Agents' Record.*

Inwood's Estate Tables.

TABLES FOR THE PURCHASING OF ESTATES, Freehold, Copyhold, or Leasehold; Annuities, Advowsons, &c., and for the Renewing of Leases held under Cathedral Churches, Colleges, or other Corporate bodies, for Terms of Years certain, and for Lives; also for Valuing Reversionary Estates, Deferred Annuities, Next Presentations, &c.: together with SMART'S Five Tables of Compound Interest, and an Extension of the same to Lower and Intermediate Rates. By W. INWOOD. 22nd Edition, with considerable Additions, and new and valuable Tables of Logarithms for the more Difficult Computations of the Interest of Money, Discount, Annuities, &c., by M. FEDOR THOMAN, of the Société Crédit Mobilier of Paris. 12mo, 8s. cloth.

"Those interested in the purchase and sale of estates, and in the adjustment of compensation cases, as well as in transactions in annuities, life insurances, &c., will find the present edition of eminent service."—*Engineering.*

"'Inwood's Tables' still maintain a most enviable reputation. The new issue has been enriched by large additional contributions by M. Fedor Thoman, whose carefully arranged Tables cannot fail to be of the utmost utility."—*Mining Journal.*

Agricultural and Tenant-Right Valuation.

THE AGRICULTURAL AND TENANT-RIGHT-VALUER'S ASSISTANT. By TOM BRIGHT, Agricultural Surveyor, Author of "The Live Stock of North Devon," &c. Crown 8vo, 3s. 6*d*. cloth. [*Just published.*

"Full of tables and examples in connection with the valuation of tenant right, estates, labour, contents, and weights of timber, and farm produce of all kinds. The book is well calculated to assist the valuer in the discharge of his duty."—*Agricultural Gazette.*

J. OGDEN AND CO. LIMITED, PRINTERS, GREAT SAFFRON HILL, E.C.

C. Tomlinson. Illustrated. 1s. 6d.

A SELECTION FROM WEA...

...AM ...
...uide to ... ...age
Machinery ... ...on.
Powis Ba... ...ds, 3s.

'RACTICAL MECHANISM, the Elements of, an
Machine Tools. By T. *Baker*, C.E.; with Additions by J.
Nasmyth, C.E. 2s. 6d.; cloth boards, 3s.

'HE BRASS FOUNDER'S MANUAL: Instructions for Modelling, Pattern-Making, Mounting, Alloying, Turning, Burnishing, Bronzing, &c. By W. Graham. 2s.; cl. bds., 2s. 6d.

IODERN WORKSHOP PRACTICE as applied to Marine, Land, and Locomotive Engines, Floating Docks, Bridges, Cranes, Shipbuilding, &c. By J. G. Winton. 3s.; cl. boards, 3s. 6d.

RON AND HEAT, exhibiting the Principles concerned in the Construction of Iron Beams, Pillars, and Girders. By J. Armour, C.E. Numerous Illustrations. 2s. 6d.; cloth boards, 3s.

'OWER IN MOTION: Horse-power Motion, Toothed-Wheel Gearing, Long and Short Driving Bands, Angular Forces. By J. Armour, C.E. With 73 Diagrams. 2s. 6d.; cloth boards, 3s.

'HE APPLICATION OF IRON TO THE CONSTRUCTION OF BRIDGES, GIRDERS, ROOFS, AND OTHER WORKS. By Francis Campin, C.E. 2s. 6d.; cl. bds., 3s.

'HE WORKMAN'S MANUAL OF ENGINEERING DRAWING. By John Maxton, Instructor in Engineering Drawing, Royal Naval College, Greenwich. 3s. 6d.; cl. bds., 4s.

'HE METALLURGY OF IRON. Containing Methods of Assay, Processes of Manufacture of Iron and Steel, &c. By H. Bauerman, F.G.S., &c. Fifth Edition, revised and enlarged. Numerous Illustrations. 5s.: cloth boards, 5s. 6d.

'OAL AND COAL MINING. By Warington W. Smyth, M.A., F.R.S., &c., Chief Inspector of the Mines of the Crown and of the Duchy of Cornwall. Sixth Edition, revised and enlarged. 3s. 6d.; cloth boards, 4s.

'HE POWER OF WATER, as applied to drive Flour Mills, and to give motion to Turbines and other Hydrostatic Engines. By J. Glynn, F.R.S., &c. 2s.; cloth boards, 2s. 6d.

HE SMITHY AND FORGE, including the Farrier's Art and Coach Smithing. By W. J. E. Crane. Second Edition, revised. Illustrated. 2s. 6d.; cloth boards, 3s.

HE SHEET-METAL WORKER'S GUIDE: for Tinsmiths, Coppersmiths, Zincworkers, &c. By W. J. E. Crane. With 94 Diagrams and Working Patterns. 1s. 6d.

RON BRIDGES OF MODERATE SPAN: The Construction and Erection. By Hamilton W. Pendred, Inspector of Ironwork to the Salford Corporation. 2s.

Spotted something that isn't quite right?

Let us know and we'll fix it for future reprints.
And as a thank you for your diligence, we'll
send you a free Sothis book of your choice.

publisher@sothis.press

This page intentionally left blank

Except of course for that line above. And this one.
And whilst we have your attention, perhaps you'd
like to follow us on Instagram @sothis.press. Go on.

Made in United States
Troutdale, OR
04/27/2026

48798090R00142